RACE
THE FINAL FRONTIER!

The TRUTH behind the MYTH

The Creation of a "Post-Racial" America
– Societal Metamorphosis
Through a Paradigm Shift in Collective Thinking

Christian Veritas

Peregrine Publius Publishing

RACE: The Final Frontier!

ISBN 978-1-7352772-0-2 (Ebook)
ISBN 978-1-7352772-1-9 (Paperback)

Published by Peregrine Publius Publishing

This book is dedicated to my family (and extended family), past and present…

Their love and support not only sustained me through the research and writing of this book, but strengthened my resolution to live a life worth living.

I bequeath my children and grandchildren this little "legacy" of words. My hope is that it might help them find their own path in life and their own truth…

CONTENTS

GENESIS OF AN IDEA

RACE: The Final Frontier!

These are the voyages of the Earthship "Race."
Its ongoing mission: to explore strange new worlds.
To seek out new life forms and new civilizations.
To *boldly* go where no one has gone before!

Recognize these words?!

Substitute "Spaceship Enterprise" and you'll certainly get it unless you weren't alive in any of the years from 1963 to the present as there are three new Star Trek movies being filmed even as I type this!

Yes, this is an intentional, blatant rip-off of the iconic introduction to the long-running television and movie series "Star Trek."

Why?

Because it fits me – and my chosen topic – to a tee!

I knew I wanted to write a book on "**Race**" for a decade and wrote **Chapter Four: The Myth of Race** back in 2010, but now that I had written most of the book, I needed a catchy title, a "*hook*" as they say!

I brainstormed a lot of ideas and heard from others. All of them were alright, but they weren't quite "cool and catchy" enough, so I went to sleep to allow my mind to work on it subconsciously…

On **Star Date 061020 @ 0333**, the title of my book hit my (not so alien) Gray Matter with the force of a laser cannon fired from the Space Ship Enterprise!

As a child of the 60s, I was enraptured by every television episode of what would become an *iconic slice of Americana:* "**Star Trek.**"

While enthralled by the original series, I was almost as pleased by the multiple iterations under different Captains. Sir Patrick Stewart brought his English Shakespearean actor "chops" to the role of Captain Picard, and it worked! But of course, Captain Kirk (played superbly by William Shatner) was the first, the original, the Man!

I wouldn't go so far to say that I am a frothing "**Trekkie**," boring all my friends with "Trekno-babble" parallels to real life because…well, because "Star Trek" just did everything best.

But I did know that I wanted to be a real Captain someday and to *fly something*, I knew not what…as long as it was aerodynamic and cool, of course. Or alternatively, to "control" something aerodynamic and awesome.

But as all dreams begin somewhere in time, we have to journey back before we can move forward, to 1968-1969 at the dawn of the **Space Age**, the first manned moon mission resulting in Neil Armstrong's iconic "Moon Walk!"

These were my formative teen high school years.

My High School was truly unique.

General William Mitchell, named after the famous General who pioneered the Air Force tactic of bombing enemy ships from the air, a feat thought at best impractical, and at worst impossible. Also known as "Billy" Mitchell, my alma mater matched my own middle name of Mitchell – which pleased me to no end.

Ultra-modern, innovative and ahead of its time, it was located 15 miles south of the prestigious U.S. Air Force Academy, the last of the Service Academies to be established in 1957. Every male at my school (women weren't admitted to the Service Academies in those days) either overtly or secretly wanted to attend the prestigious "fly boys" University, the "Blue Zoo" or the "Aluminum Womb" as it was fondly - or disparagingly known - to insiders.

As such, it was written up in "Time" magazine. But what made the school unique?

Well, in addition to being designed like a college campus with college curricula, with class periods spaced out, winning football, basketball and track teams, our mascot the Mitchell Marauder, and the coolest pep rallies, it also had two swimming pools, one for each gender. That was a massive luxury in that day and age!

Academic study was strongly encouraged in various "Resource Centers." By "strongly encouraged," I mean they had nerds assigned as "Hall Monitors" to clear the riffraff and smokers out of the bathrooms and funnel them into one of six approved "Resource Centers" where they could be appropriately supervised by adults.

But the key to its uniqueness was *none* of that.

It was the **Scott Carpenter Planetarium**, dedicated by an actual American Astronaut Superhero Scott Carpenter. That was our claim to fame!

After graduating from General William Mitchell High School in 1969, I applied to and was accepted to the **United States Air Force Academy** (USAFA).

It was there that my real love for everything space (**The Final Frontier**) was set to "after burner," to use a fighter jet pilot term.

Not only did we take such courses as "Aerodynamics." "Astrodynamics," watch flyovers of all manner of inspiring Air Force and Army aircraft over our heads just walking to class, marching in a parade or walking to Mitchell Dining Hall for a Football Pep Rally, performing experiments in clean labs and in the world-class "Wind Tunnel…"

But one other thing: USAFA had its own **Planetarium**, bigger, badder, better than the one at my former high school – and it wasn't my first time studying the stars there either. From 1958-1960, at age 7-9, my father taught French in the Language Arts Department at the Academy, and we loved to attend USAFA Planetarium shows as a family.

So my exposure to this institution was a long one…and

it doesn't really end, with five-year Reunions with Classmates (Class of 1974, 8th Squadron), fundraisers, etc. In fact, as a USAFA graduate, my wife and I are entitled to a few square meters of the **Air Force Academy Cemetery** – both a comforting and disturbing thought at the same time.

My Air Force career was primarily associated with the **NORAD**, the North American Air Defense Complex southwest of Colorado Springs. While never physically assigned there, I held a Top Secret Clearance, working five years at the associated NORAD Regional Control Centers at Luke AFB and then 60 stories under a granite mountain at Canadian Forces Base North Bay! My job as an Air Weapons Control Director was to lead a team of ten men in "scrambling" and "controlling" interceptor jets to counter Soviet nuclear-armed bombers testing the "Air Defense Identification Zone" or ADIZ encompassing the United States and Canada.

My later civilian career included four years at a **Space Launch Complex** at **Vandenberg AFB** planning the space launches of Air Force Peacekeepers ICBMs! I watched scores of these launches – up close and personal – adjacent to a beach in Central California. None carried actual nuclear warheads, of course. This was a Research, Development, Test and Evaluation Program.

So there's your answer to how I came up with my cool and catchy book title! I was and still am an Astronomy buff, a "Trekkie," an Air Force Academy graduate and a "**Space Cadet!**" if you will.

BUT WHY IS "RACE" THE "FINAL FRONTIER?"

The United States of America began as a vast land,

occupied by no human, then 15,000 years ago, a small band of humans, a nomadic group, migrated over time across the Great Land Bridge (now submerged under the Bering Sea between Alaska and Asia) and into what is today the North American continent.

Once settled in these new lands, they then acquired the title of indigenous peoples, Native Americans, if you will, but of course, that title had no meaning until America itself first came to be.

Those people continued their migration throughout the Americas (North, Central and South America) for an additional 3,000 years, adapting to the new geography, climate and challenges they encountered, until they occupied all of the Americas from Alaska to Tierra del Fuego on the southern tip of South America. They became the northern Inuit, the numerous tribes we know within the contiguous 48 states as the Sioux, Cherokee, Algonquin, Navaho, Apache, Hopi, and countless others, and in what is Mexico, Central and South America today, the Aztec, Mayan and Incan tribes.

Those of us who are citizens of the Americas known collectively as these "United States of America" and who can trace our recent genealogy to Europe know that our ancestors came to these same lands far later. They were introduced to the New World in 1492 as any school boy or girl knows, but eventually permanently settled in relatively small numbers along the Atlantic seaboard in the 1600s, and despite the multinational interests of Spain, France and others, eventually became the 13 original American colonies of Great Britain.

Through the tumultuous years of the American Revolutionary War, our colonies became a new nation (under

God, with liberty and justice for all), but one which still primarily occupied only the Atlantic seaboard from Maine to Florida.

Everything west of this point was considered the "**Frontier**." In time, through the drive of her people to fulfill the country's "manifest destiny" to extend our borders from "sea to shining sea," that frontier was explored and claimed until the Pacific Ocean defined the western border of the "continental" United States, consisting of the 48 contiguous states, eventually to be joined by Alaska and Hawaii (hence 50 states) and our U.S. Territories.

So in the physical sense, the "Frontier" had been explored and settled.

However, we would soon learn that another great chasm divided us: **slavery**. From 1861-1865, roughly 400,000 men would die to end the practice of slavery. Of course historians tell us and rightfully so that there were other forces at play, among them a debate over states' rights, but no other issue was as divisive as that of slavery.

So "**race**" – what you will come to understand by the reading of this book is a false social construct and not a biological reality at all – was then, and remains today a divisive and destructive notion I seek to eliminate entirely. In the process, the associated notions of "racism" and "systemic racism" will also hopefully fall from grace.

"Race" was then – and is today – our last "Frontier," as challenging as any that came before it, but explore it and conquer it we will.

A "**Frontier**" is also one of knowledge: the extreme limit of understanding or achievement, and if science is successful, it provides a commonly-accepted basis for understanding so we can finally achieve "**unity**" in our belief system.

We have not yet achieved that milestone – and it is in this spirit I wrote this book.

CHAPTER 1

INTRODUCTION

The Early, Formative Years of my own Concept of "Race"

In 1974, I selected a Senior Elective course entitled "**Minority Studies**." I attended class with several black Cadets, who obviously brought their own perspective to the classroom.

But I had spent three years in the 60s experiencing the "**Civil Rights Era**" in Montgomery and Selma, Alabama, had met the infamous Governor, George Wallace and went to Sidney Lanier High School for a year with his children before transferring to General William Mitchell HS in Colorado Springs, Colorado.

As you might imagine, we had respectful – we were in training to lead men in combat and in the air, after all, but still *vigorous and enlightening debates ensued on the topic of RACE* for months on end.

Concurrently, I had enrolled and completed two courses in "**Anthropological Studies**." To say this opened my eyes to a new understanding of myself and my fellow man (as in the inclusive definition of mankind, which encompasses both male and female) through a study of my recent and distant ancestral roots would be an understatement.

I also graduated from the University of Southern California in 1982, obtaining a Master of Sciences degree. For my academic performance, I received the Tactical Air Command Distinguished Scholastic Achievement Medal. We were required to write 12 thesis papers on various topics, and one of the papers I submitted was "The Myth of Race" upon which this book is based.

These combined academic and somewhat personal experiences offered me an opportunity to form an adult perspective and opinion on the topic, and although, I certainly didn't recognize it at the time (that took place in 1974, and I wrote this in 2020!), it was nonetheless the genesis of THIS book!

[This is not to say that I had no foundation in the topic, as my upbringing and childhood experiences had already formed within my young mind a belief system in the equality of mankind. My childhood experiences and influences are revealed in Chapter Two.]

Follow that with 45 years of hobby reading in the sciences, including Anthropology, watching years of ***Discovery Channel*** and reading Science magazines. Discovery Channel did several series on race and culture though the 80s and 90s – and I watched them *all.* I consumed the **Special Issue on "Race" published by National Geographic** in April 2018.

So enough about me and my qualifications to write on the topic of RACE…

It's time to taxi out onto the runway of life (yes, I love aerospace references!) – and take off!

No judgment will be made by me as to who you were before you read this book. No premature assessment of what kind of human being you might soon become after you finish and set this book on your "virtual bookshelf."

Perhaps you will join me in transcending this mundane plane of existence to a higher calling in the "Post-Racial" America and world we hope to create…

Now *turn the page* to begin your personal voyage of discovery!

CHAPTER 2

THE JOURNEY BEGINS!

[Iconic Movie Scenes / Red Pill / Blue Pill / The Rabbit Hole / Run toward the Light!]

"We must not cease from exploration.
And the end of all our exploring
Will be to arrive where we started
And to know the place for the first time!

Through the unknown, remembered gate
When the last of earth left to discover
Is that which was **the beginning**;
At **the source** of the longest river

The voice of the hidden waterfall
And the children in the apple tree
Not known, because not looked for
But heard, half-heard, in the stillness
Between two waves of the sea {of humanity}

~ T.S. Eliot

I can see now that the gradual PATH I myself took – "the one less traveled" – and the *accelerated path* I am inviting you to take with me…perhaps better thought of as a ***journey toward knowledge…is metaphorically identical to the famous scene in the movie called "The Matrix."***

This very night (June 7, the day my mother passed away and entered into Heaven), I woke at 3:33 A.M. as if from a fevered COVID-19 dream, sweating profusely, clutching the damp sheets in a balled fist…

I had experienced an epiphany, a sudden moment of revelation and insight of what I am compelled to do…what I must do…to reveal the **truth** of what I know to others regardless of the consequences.

Throwing caution to the winds…have I always been so cavalier and rebellious?

No, I would say not.

However, I have never been sufficiently obedient either, not even in the structured, rigorous environment of a national military academy in the early '70s, an institution I have both loved and despised at times, but with age, we mellow, and I have come to think of that place and my experiences there fondly…and in returning to it for my 45th Reunion in November 2019, perhaps came **to know the place for the first time.**

I have never kneeled repentantly before any person.

I have never been mastered by anyone, but would be more than happy to embrace every single human being on Earth as my undisputed equal in the eyes of God – and in conformance with that hallowed document, the U.S. Constitution (and its 27 inspired Amendments; 10 of which form "The Bill of Rights"). We all read them once in school – and then promptly forgot them.

With no exceptions other than God and Jesus Christ Our Lord and Savior, I am my own master. I am willing to be a figurative "slave," if you will, to only one master: the TRUTH. At least the truth as I sincerely believe it to be. Not someone else's truth, but my own.

Whether I am right or wrong in my beliefs is someone else's task to ferret out for themselves. I can't concern myself with that right now. I must simply be willing to reveal a truth that so few, I have come to realize over the years, know.

Not only do they not KNOW the TRUTH; they are in fact entirely – and unfortunately not blissfully – unaware of its existence.

As you read through this book, you will find yourself feeling as if you are chasing the proverbial "White Rabbit"… that you are like inquisitive Alice in Lewis Carroll's (aka Charles Dodgson) "Alice in Wonderland" falling through the "Rabbit Hole"…into a strange, surreal world not of your making, yet at the same time feeling as if you are being "enlightened."

Whether you actually achieve the "Brahman" level in your 3rd Reincarnation of yourself, perhaps reaching your own "**racial nirvana**," is an unknown at this early juncture. But one thing is certain: You are already on your way to a new destination!

My fervent wish for you is that you will come to see beyond the illusion of your current reality, to allow a purer, more truthful reality to seep into your consciousness. Not a cultish vision, God forbid, but one of truth as revealed by a love for all of humanity.

That a thick veil that has obscured your vision all your life is gradually being lifted, allowing you to return to the same place...and to know the TRUTH for the first time!

The Hollywood movie scene that formed such a perfect metaphor of what this book symbolizes...that came to me in that fevered dream this past night, was the following one.

I have taken great pain to capture it for you verbatim – and to paint a mental image by describing the scene as it unfolded visually on screen...

"**The Matrix,**" 1999

Scene: **RED PILL** versus **BLUE PILL**

Setting: Two men, one black man (Morpheus), older (wiser, we wonder), wearing impenetrable mirrored sunglasses; one white man, young, dressed in black, his hands perched defensively on the arms of the chair, ready to spring away if need be, but transfixed by his own nagging curiosity. They sit alone in a darkened room, thunder rolling ominously outside, facing each other in two large, high-backed vintage leather chairs.

Morpheus: Do you want to know what IT is?

The Matrix is everywhere; it is all around us. Even now in this very room. You can see it when you look out your window or when you turn on your television. You can feel it when you go to work, when you go to church, when you pay your taxes...

It is the world that has been pulled over your eyes to blind you from the TRUTH.

Neo leans in toward Morpheus expectantly.

Neo: **What truth?**

Morpheus leans forward to close the gap between them...

Morpheus: That you are a slave, Neo. Like everyone else, you were born into bondage, born into a prison that you cannot smell or taste or touch...**a prison for your mind**.

Thunder rolls in the distance...

Morpheus sits back in his leather chair and sighs...

Unfortunately, no one can be *told* what the Matrix is. You have to see it for yourself.

Morpheus opens a thin, black pill box...he removes two

identical pills and sets the pill box on the table separating them…only a half-full glass of water on the table.

The pills are identical except for their **color.** One pill is **red** and the other is **blue**. He places one pill in each of his outstretched hands, palms up, ready for the selection, ready for the taking…

Morpheus (speaking in an ominous tone): This is your last chance…after this, there is no turning back…

You take the **blue pill** ("left" hand), the story ends, you wake up in your bed, believe whatever you want to believe…

You take the **red pill** ("right" hand)…you stay in Wonderland…and I show you how deep the rabbit hole goes.

Neo reaches out for the **red pill**. His face is reflected back to him by Morpheus's mirrored lenses.

Morpheus forces him to pause…to take a moment to reconsider.

Morpheus: **Remember: All I am offering you is The Truth…nothing more**.

Neo reaches out, takes the **red pill**, places it in his mouth, accepts the glass of water and washes it down without hesitation.

Fade to black…

* * *

Cut to "**Alice in Wonderland**" (in my dream) –

Alice, in her long robin's egg blue dress and white stockings, runs down a forest path after a large White Rabbit, stops at a tree stump, shouts out "Helloooo" into the dark hollow at its base. Suddenly, she trips over a root, *falls into the rabbit's hole*, falling endlessly down an earthen tunnel resembling the interior of a tornado, with swirling debris from a past life...*and lands in a strange, new world.*

I would even throw in Aldous Huxley's "Brave New World" (1932) because...

...brave you must be to take the red pill because not all of your family, friends and colleagues will.

But there we will divert from the intent of Huxley's nightmarish vision of a future society.

Instead, **I offer you a vision of a far better society**.

The "happily ever after" ending will be to become "one" in spirit...and also you will have "won" in reality. Its ultimate reality will only fully manifest in Heaven, but perhaps we can create a lesser version on Earth and not have to wait quite so long...

So what is this I offer?

Well, dear friend, it can neither be described in a single sentence nor absorbed fully in a lifetime, but the *essence* of this new TRUTH is this:

RACE DOES NOT EXIST! It never has; and it never will!

Whoa, hold on, I imagine you thinking.

Is he serious?! Or is he insane?!

Did I really just spend $9.99 only to learn that something that obviously is not true... is after all, really true?!

Mind officially blown!

You might even feel a small betrayal, bile rising into your mouth, anger rising in your chest…

I do sincerely apologize, but yes, that's exactly correct.

You spent your single "Alexander Hamilton" Federal Reserve Note *to entirely upend your outdated thinking, your valued belief system, your fatally flawed "**construct**" about "**race**."*

Isn't this what you expected? Perhaps secretly hoped for?

Was the status quo of divisiveness and hate really your "happy place"? Are you really going to miss it? I don't think so!

All I promised you was the **truth**, the whole truth and nothing but the truth, so help me God – and we *will* get there!

Patience, dear Reader!

Think of me as your guide leading you on a private tour (usually a lot more costly) of your heart and soul. Follow me on this journey: I will lead you to a new, potentially life-changing perspective! Call me crazy, but I think that has value – even beyond a ten spot!

Wait, Christian, please slow down, you implore…
You lost me at "construct."
What in God's name is a construct?!

Construct (definition) – An idea or theory containing various conceptual elements, typically one considered to be **subjective** and not based on **empirical** evidence.

In other words, **what you have always believed to be the truth was in fact subjective and never was based on empirical evidence**. It really is that simple, yet complex.

But do not dismay!

In future Chapters, as concisely as I can, I will build you a **new, improved construct**, *one that will hopefully forever alter your perception of race in a positive way.*

In fact, I intend to entirely obliterate that divisive notion taught to you or learned as you grew up! In its place, I will fill that void with no cultish concept, but a promise only that I will reveal to you the TRUTH as I sincerely believe it to be. No brainwashing. It is entirely up to you whether you accept it or not.

Remember: All I am offering you is *The Truth*... nothing more.

But no worries, friend!

Of course I won't leave you in the dark much longer: **The whole point of this treatise is to lead you to the light!**

And not years or months or even days hence...delaying the inevitable would be too dangerous...but now! Now through the tried-and-proven mechanism of metaphor...

As you read through the upcoming movie scene from "Poltergeist," pay particular attention to those words which I have highlighted in **bold font**. They tie directly into **The Message** I am trying to convey!

Our guide through this compelling and frightening scene is the psychic / medium / **Clairvoyant** Tangina Barrons. That part is played by the wonderful actress Zelda May Rubinstein, may her soul rest in eternal peace! Zelda – although not widely known – was a "Human Rights Activist," which aside from my liking the movie and her acting in it, was the real reason why I chose her to be our guide toward the Truth.

When you see that the word "Clairvoyant" Tangina is speaking in the scene, consider for the purposes of ***this* journey of metamorphosis** I am leading you on that **I** am the **clairvoyant** – and appropriately so, as I myself have the God-given gift of insight and the perception of future events.

I am the Clairvoyant in this scene.

Not like a prophet, psychic or palm reader!

No, I would prefer to think of myself more like a male version of the ancient **Greek Oracles, chosen by the Gods (or in my case God) through which divine advice and truth would be revealed!** In *that* analogy, **you** would be the Pilgrim who has made his or her way from some "far off place": (i.e. some people's wrong thinking: their engrained, outmoded concept of "race")

But no, In reality, my gift of "insight" really is only the by-product of 70 years of study, thought and life experience culminating in my (Star Trekkian) laser-beam focus on one topic: **The Myth of Race**.

I have happily spread the word person by person for decades, but I grew frustrated with the size of my audience. Hence, this book…

So without further ado, let's launch into our brief flight of fancy toward another frontier, the boundary between life and "after life" (in our **metaphor: it would represent the boundary between the current thinking on "race" toward my dream of a "post racial" era** when all our earthly misconceptions evaporate into a mist, never to be resurrected)…

And what lessons can be gleaned through self-enlightenment…

Screenplay Script from
"**Poltergeist**," Horror - 1992:

***Scene*: Run toward the light!**

Setting: Interior foyer of a house built on an old, forgotten cemetery, which has stirred restless spirits who are neither of this world nor yet of the next.

Diane and Steven Freeling, suburban California parents of a little girl, Carol Anne, who has gone missing, have invited a "**Clairvoyant**" Little Person by the name of Tangina Barrons *to find and bring her back into the real world*.

Their daughter Carol Anne has been abducted by a single malevolent force with the ominous moniker "**The Beast**" among other **harmless, restless spirits who are temporarily caught between *two worlds*...**

[Throughout this book, **if I reference "*The Beast*," you will understand it to act as a metaphor for the *scourge of "systemic racism"* that some people insist exists, but which does not, in that from its inception, it was fabricated from whole cloth of a lie!**]

As you will come to understand by the end of Chapter Four: "The Myth of Race."

Clairvoyant Tangina's key monologue: There is no death! **There is only a transition to a different sphere of consciousness**. Carol Ann is not like those she's with. She's a living presence in their spiritual, earth-bound plane.

***They're attracted to the one thing about her that's different from themselves: her life force! It is very strong... it gives off its own illumination*. It is the light that applies life**...and memory of love...and home...and earthly pleasures. **Something they desperately desire...but can't have any**

more. Right now, she's the closest thing to that...and that is a terrible distraction from the *real light that has finally come for them.*

Do you understand me?

These souls – who, for whatever reason, are not at rest, are also not aware that they have passed on. They're not part of consciousness as we know it. They linger in a perpetual dream state...a nightmare from which they cannot wake. (***Every time they turn on the news or check their social media!***)

Inside the spectral light is Salvation...a window to the next plane. They must pass through this membrane...their friends are waiting to guide them to new destinies. (***Just as I will, by the end of this book, help guide you to your new destiny. The passages in bold font will come into clear focus in time.***)

Carol Anne must help them cross over...and she will only hear her mother's voice.

Now, hold onto yourselves... [Gets down on her knees to speak with them eye-to-eye at their seated level]

There's one more thing: **A terrible presence is in there with her...so much rage...so much betrayal...I've never sensed anything like it**...I don't know what hovers over this house, but they want to take your daughter away from you.

It keeps Carol Anne very close to it...and **away from the Spectral Light. It lies to her.**

It says things only a child can understand. It's been using her to restrain the others. To her, it simply is another child. **To us, it is *The Beast!***

Long pause…

Clairvoyant (with strong resolve in her voice): Now let's go get your daughter!

Author's voice (with strong resolve): **Now let's go find the Truth!**

* * *

Clairvoyant: Now clear your mind…it knows what scares you…it has from the beginning. Don't give it any help…it knows too much already.

Salvation is waiting for you in the light! Run toward the light!

Clairvoyant (loudly): **Now open the door!**

CHAPTER 3

MY PERSONAL AWAKENING TO THE FALSE NOTION OF "RACE"

Author: That's right! **Open the door!**

Throughout this book, **if I reference "*The Beast*," you will understand it to act as a *metaphor* for the *scourge of "systemic racism"* that has total control over your life, that prevents you from participating, getting ahead or which creates in you *any* self-limiting beliefs.**

Some insist it exists, but you will come to understand over the course of the book that it does not, in that from its inception, it was fabricated from whole cloth of a lie!

Be patient. This will all become abundantly clear in **Chapter Four: THE MYTH OF RACE!**

When we were children, it was our parents' job to coddle and comfort us, hold our hand gently, feed us baby pablum, one tiny, silver spoonful at a time, wipe our chubby, sweet little faces and tell us bedtime stories and fairytales!

As adults, we have accepted adult responsibilities. All of us must now release our childhood stories about **The Beast**!

Don't fear **The Beast**! Wake now to a more optimistic future.

I will only reveal what I have discovered through years of study and experience and have come to believe confidently is the truth.

Just the **TRUTH**. No agenda!

O.K. Alright! Yes, I admit I *do* have an agenda. But I will *never* veer from the truth as I know it to serve a narrative or an agenda.

I am a Conservative Christian who supports both the letter and more importantly, the spirit of the Constitution of the United States. As have millions of Americans, especially those in the military and new American citizens, I have sworn to support and defend the Constitution against all enemies foreign and domestic.

I try to live by Judeo-Christian ethics.

As such, I love you enough, indeed so much that I will bare my soul to you if the result is that I can help you to you face the TRUTH and *to run toward the Light*! You will come to understand what I mean by this in time…

The **good news** (like in the Christian Gospel) is that the truth is beautiful, not harmful, not damaging to your psyche, but a palliative to your pain.

You have always felt the pain from the moment you "learned" that there were so-called different "races." You didn't know why this hurt you, only that it did.

Thankfully, I was raised in a loving home. Without exaggeration, I can say my childhood was idyllic. If you were raised in a less than idyllic household, unfortunately, you might have been taught about "differences" and that these so-called "differences" would separate you from people you had befriended, possibly even loved, because they were not of your flock, but another. "Johnny, remember: Birds of a feather flock together!"

Perhaps your little friend Kenton was no longer allowed to come over for sleepovers. This hurt you…you missed your friend…but nobody would tell you why you weren't allowed to play with Kenton any more.

I searched my personal history nostalgically to attempt to discover the genesis of my own racial awareness. If it was not formed in our loving, accepting home, then where…and when?

Flashback to my experiences for four years living and studying in France from 1960-1964, traveling extensively throughout Europe with my Air Force Officer father, my mother, older sister and brother. These were formative years indeed!

I wish I could transport you, dear reader, via a Time Machine to the exciting adventure that traveling through Europe in the 1960s was. With countries the size of U.S. States, and un-forbidding border crossings (like today), you could quickly find yourself immersed into an entirely different world!

Different language, cuisine, culture, history, world outlook, currency. Different people who looked different than you and from the people you had just spent a joyous time together with the previous week or day!

Acceptance of the apparent "difference" between people became second nature to me. In fact, most of the joy I felt on our international travels came from the diversity of the people and of the experience we had in each postcard perfect locale.

As the Aide-de-Camp, interpreter and part-time pilot for a Five-Star French General, working within AFCENT (Allied Forces Central Europe), my father was extremely fortunate to have his office located inside the Palace of Fontainebleau, a favorite retreat of Francois I and Napoleon!

My favorite story – of which I have many – associated with Napoleon's Chateau was when my mother dropped me off after spending the day at L'Ecole International at my father's office.

My father hadn't completed his work, so he directed me down the hall, suggesting that rather than be bored for the next hour, I should instead, at age 11, give myself an unauthorized, unguided tour of the Palace. It was not open to tourists on that day and hour, so delighted, off I ran.

It was like stepping into an historical period piece: ornate, gilded sculptures, priceless treasures everywhere, Renaissance oil paintings hanging within elaborate, carved frames. Twenty minutes into my self-guided tour, I began to tire, decided to slip under a drooping, burgundy velour rope intended to keep prying tourists off the furniture and from absconding with a fortune in artifacts!

Later, my father found me sound asleep in one of the bedrooms, not just any of the scores of guest bedrooms, but the very canopied marital bed allegedly belonging to Napoleon himself and his lovely bride Josephine, her beauty immortalized for eternity in a life-sized marble sculpture, her reclining form posed, partially nude, but demurely draped.

My father chose not to scold me for going a little further afield than he might have liked, and instead we shared a good laugh over the possibility that nobody had ever slept in that bed other than Napoleon, Josephine…and me!

We learned the overseas military way of life, a life nearly devoid of racial distinction. Racial discrimination, were it to rear its ugly head, could be career ending in the military. There was no room for race on the battlefield or in combat. Leaders were schooled at West Point and at other military institutions to be "color blind." The Military, Sports and Academia were the first bastions and institutions post World War II to create racial equality to the extent that it could exist within the larger and less accepting…and in some ways, the more dangerous outside world.

So no, not there in post-World War II Europe did I find the genesis of my own "racial" awareness. No, for that, we need to move forward from this period, but still long ago on the timeline of my life…56 years ago to be exact!

Flashback to 1964-1967:

These were not the years of origin for "The Civil Rights Movement," which began just after World War II, but in the mid-60's, the movement was alive, well, and still in full swing in the Deep South.

Jim Crow Laws were supposedly signed out of existence in 1964 when President Lyndon B. Johnson signed the Civil Rights Act. But that was the legal end. The pernicious effect of Jim Crow prevailed and was still everywhere in evidence: "Whites Only" bathrooms, diner counters, drinking fountains, country clubs…

Violence surrounded ALL of these events.

Ever see the wonderful movie, "**The Help**"? If so, then you already know the milieu into which I was thrust – out of France and into the Civil Rights Movement in Montgomery, Alabama in the summer of '64. In some ways, it was like living inside that movie!

But fortunately for my family…and for my 13-15 year old self, *not* exactly.

We lived in base housing on Maxwell Air Force Base, a microcosm of U.S. military life everywhere, an island, a safe refuge from the "racial strife" fomenting outside its gates.

It was an environment we military members and dependents all knew, loved…and most importantly, understood.

On every base our father was ever assigned, we found the same comforting environs: an "Officers' Club" (and Non-Commissioned Officers' (NCO) Club; fine family dining (for *all* families); a friendly, accepting "all races" bar for the adults; a bowling alley; a non-denominational church; a movie theater; an all-you-can-eat "Mess Hall" for soldiers; a Commissary and a Base Exchange (BX).

The BX provided discounted consumer goods and services for active duty, reserve, national guard and retired members of the U.S. Uniformed Services, their dependents (spouse and children), and starting in 2017, all honorably discharged veterans. It was a perk and we knew it.

Always a "Commissary" (a store that sells food and supplies to the personnel on a military post). Another money-saving perk we took full advantage of.

Always a fenced compound, a *guarded* front gate, a *guarded* back gate, a perimeter road where military dependent teens attended "driver's education classes and learned to drive a car from their parents without being exposed to the larger, "different world" outside that perimeter.

So no, *not* there in our protective womb did I find the genesis of my own "racial" awareness.

In my own case, my so-called "racial" awareness began outside the gates of Maxwell AFB, at Baldwin Junior High School, a brick multi-story structure with broken windows, no gym, no swimming pool, no showers to change after P.E.

I could not relate to the first boy I met who had never been outside his own hometown except to accompany his father hunting, but only into the next county! Even at this young age, I had already lived in Europe, Japan, China, seen and experienced a lot within those worlds.

Instead, I befriended a "black" boy named Franklin, my 8th grade home room and 1st Period buddy – the first of his "kind" ever to attend Baldwin Junior High School in Montgomery, Alabama in 1964. But he was a military dependent and so we were instantly fast friends.

One day Franklin was confronted, then attacked on the playground by three bullies. I was outraged and blindly struck out with the only weapon I possessed at the time (before I found my voice): my fist. I punched the leader square on the chin, and he went down – hard. His little "KKK henchmen" – as I came to think of them in time, dispersed in fear as the "school in session" bell rang loudly.

I thought: ***Why do I feel really good, yet bad at the same time?"***

Another little epiphany, perhaps my first: **So *this* was the genesis of my so-called "racial" awareness!**

In time, I would also realize that those bullies were just the product of their upbringing, that none of us is born with hatred for those different than ourselves, but rather that behavior is learned, usually at home first, then reinforced by peer pressure and society as a whole.

As I walked back inside the school, my 13-year old heart beating rapidly, pounding in my chest, I noticed in my peripheral vision a solitary adult figure standing on the retaining wall adjacent to the school recess yard. I recognized him. He was the School Principal whose name I long ago forgot.

I suddenly realized that I had noticed him earlier, even before the horrid confrontation. He had witnessed the schoolyard fight, yet did not intervene. He had witnessed my act of violence – but an act born of love for my "black" friend – and he had not reacted. He had done nothing!

I struggled then (and ever since) to understand why. Was he siding with them or was he siding with me? Or was he held fast in place by the sudden, violent episode he himself didn't know how to handle even as an adult?

I will never know the answer, but I would like to think that he was caught up in a tide toward racial integration, equality and progress, had a conscience and wanted to be on the right side of history.

What neither the School Principal nor I knew then was that, in time, **I would come to question every aspect of "race," including its very existence.**

So now that you understand the *journey* upon which you are to embark, are hopefully eager to "have the veil lifted from your eyes so that you might see again anew, to receive the gift of second sight"...the supposed ability to perceive future or distant events, perhaps even "clairvoyance" (like Tangina Barrons in "Poltergeist") from the Latin, meaning "clear seeing or vision" or "insight."

It's time to select the **Red Pill** or the **Blue Pill**...

The Author (in an ominous tone): **This is your last chance...after this, there is no turning back...**

You take the **Blue Pill** ("left" hand), the story ends, you wake up in your bed (not Napoleon's), believe whatever you want to believe...

You take the **Red Pill** ("right" hand)...you stay in Wonderland...and I show you just how deep this rabbit hole goes.

If you are brave and daring enough to select the **Red Pill, turn the page!**

CHAPTER 4

THE MYTH OF RACE

"There is a great new work before us, which is to replace with true knowledge the past ignorance that has destroyed human minds!

We will construct unity in a world which has been brutally torn apart by *false divisions of race...*

We will heal with unconditional love those souls whose hearts have been disfigured by hatred and loneliness."

~ Aberjhani, Songs from the Black Skylark

Yes, I boldly confess: my mind is alert, my heart is receptive. I will never be the same. Oh, sorry...sleep deprivation has me channeling Joel Osteen! But honestly, I can't think of a better introduction to this Chapter.

I can't fully capture the contents of all the television programming I've seen on the subject nor condense all the course work I have taken nor published articles on the topic

I have read. Nor can I present the material as well, especially lacking the visual graphics of the Discovery Channel and the credibility that the numerous experts in their respective academic disciplines interviewed for these programs cumulatively possess.

My intent is to shed some needed light on the ill-conceived, extremely long-lived, but misguided and divisive notion and false social construct of "race" (used by some to justify the most horrible atrocities in history).

Over a span of years, the Discovery Channel presented a variety of programs on the origins of man, detailing the work of paleo-anthropologists, paleo-archaeologists, population geneticists and biologists, pinpointing the origins of man, tracing the movements and routes of the human race out of Africa, etc.

In April 2018, National Geographic would publish a Special Issue on "Race."

The conclusions presented were derived from prior, painstaking studies by thousands of scientists of the skeletal remains of our earliest ancestors, by the complex undertaking of drawing blood samples from tens of thousands of living representatives of every culture in the world, studying the genetic make-up of their DNA, etc.

Most people "know," or should – especially if they are either scientists or literalists when it comes to the Bible (specifically Genesis) - that we all share a common ancestry. But relatively few have also come to the further conclusion that "**there is no such thing as race.**"

In fact, most people are completely incredulous when confronted with this fact.

People often confuse or use the term race and ethnicity interchangeably. This is incorrect. Ethnicity is defined as "the fact or state of belonging to a social group that has a common national or cultural tradition." Ethnicity exists for reasons that will become increasingly apparent as you read further. However, "race" does not.

I don't use the word "theory" because it is an almost universally-accepted truth in the scientific community, a confidence that has developed over time in the light of the overwhelming accumulation of irrefutable, peer-reviewed scientific data. One would be hard pressed to find a reputable anthropologist (or scientist in related fields) that believes otherwise.

Let me first lay out a series of accepted facts, which when taken in combination, and illuminated through various examples, should help make it clear why there are no clear divisions, no biological categories scientists recognize as constituting different "races" of human beings.

1. Every person has or had a mother (and father, of course, but simplifying here). Your grandmother is your mother once removed. If she didn't exist, you would not exist. No matter how many generations one considers, and how many "greats" you add to the title "grandmother," *they are all individually responsible for your existence and they have all individually contributed to your genetic code.* (In other words, every one of these individuals is your "**mother**," whether you gestated in her womb and/or she "parented" or raised

you or not.)

2. Today, mankind covers most of the land surface of the Earth, but that was not always the case. *As the remains of older and older modern humans (then their hominid predecessors) are discovered, the areas in which they are found become more and more localized.*
3. Today, 7.5 billion people live on the Earth. However, *as we look further and further back in time at human populations, the numbers eventually dwindle to relatively few individuals.*
4. Your blood and/or saliva can be drawn, and from the DNA contained therein, various genetic markers can be pinpointed, tying you to a specific lineage. (By now, most of us are familiar with Ancestry.com, 23andMe.com, etc., and might have even had a saliva sample provide us our recent roots).
5. All living organisms have a capacity to adapt to their environment to some extent. If this were not true, flora and fauna would become extinct merely by migrating to a different area or experiencing a change in the climate.

Ultimately, it has been determined that every person on Earth originated from the same geographic areas – generally the northern end of the great African Rift Valley in northeastern Africa, bounded today within the modern day country of Ethiopia, from an area bounded by modern-day Morocco and in the south, Botswana.

Population geneticists have been able to trace all living beings to a common "mother" and in a separate study, to a single man or "father."

To be clear, *scientists were able to genetically trace every single person on Earth living today to a female who lived in Africa in the distant past. A single male specifically contributed the "Y" chromosome portion of his DNA to all male successors.* However, many other genetic traits were contributed from other males from the same region.

In the same area as the excavation sites of the oldest known human remains, one scientist discovered a set of young adult human female footprints miraculously preserved in hardened sediment along the banks of what had been an ancient river.

From various studies, including that of the partial skeletal remains of the woman's "tribe" in the area known to be the birthplace of mankind (i.e. where the oldest remains ever found are located), the dating from the layers of soil and rock that act as a calendar back to the origins of mankind, etc., the scientist instantly recognized the relevance of his discovery.

He removed his boots and socks and fit his larger foot into the woman's footprint. An unexpected wave of emotion overcame him as he realized that he was standing in the footprint of one of his many "mothers." In fact, he was standing in the footprint of the common Proto "mother" we all share. Yes, certainly a grandmother with a huge number of "greats" in front of that title, but that notwithstanding, the same "mother" of us all – and **our "mother" was African**.

She would not in effect actually be the "Eve" we know from the Bible only because although she is our common root, she was not the first woman to live. Hence, the name "Mitochondrial Eve" is used to reference her.

She was a woman who lived approximately 200,000 years ago. The timeframe has also been set at about 200,000 years for the appearance of modern day man on this planet. We were eventually to be labeled by family / genus / species as "homo sapiens sapiens," the Latin title applied to every living man, woman and child alive today.

The migrations of subsets of our ancestors in two primary waves out of Africa falls generally within a period of only 100,000, then 50,000-60,000 years ago.

[Not a part of this discussion is that what constitutes "modern day man" versus "modern day behavior"; they are not the same. All the things we associate with modern man today, the non-nomadic practice of settling in a given area, the development of tools, trade, etc. came far later in history than the date of our origins.]

Scientists first theorized, then all but determined with certainty through DNA tracing, that *there were specific migrations of mankind out of Africa in multiple "waves"* spanning tens of thousands of years.

As such, Africa has been called the "birthplace and cradle of mankind," and with the convergence and subsequent outgoings of mankind first to, then from Central Asia, this region has become known as mankind's "nursery." Similarly, the ancient Near East has been called "the cradle of *civilization*" (evidenced by the earliest human settlements).

From there, some groups went west to Europe, and some east to the rest of Asia, some skirted the coasts, and some headed north to Siberia and the Arctic. From there, they traveled over an enormous Siberian land bridge across the Bering Strait called Beringia, and these few people (astoundingly, possibly as few as 10-20 people total!) eventually became the hundreds of native tribes of North, Central and South America.

Be they Inuit, Sioux, Cherokee, Hopi, Apache, Navaho, Inca, Maya or Aztec, they all trace their common roots to the few people that came across the massive land bridge (that exists, but lies under the Bering Sea today), and prior to that, from Siberia / northeastern Asia, and prior to that, from the nursery of Central Asia, all of whom came from their birthplace in northeastern Africa.

Due to the isolation of populations in different regions that presented different "environmental pressures": climate, latitude, differing intensities and directness of sunlight, differing temperatures, differing food sources, etc., groups of humans adapted physiologically in various ways, not to mention varying their language and customs.

For example, peoples of the frigid far north, over multiple generations, developed stouter bodies, shorter arms and legs, shorter fingers, etc. This was a natural adaptation to the extreme cold and to the biological survival mechanism inherent to all humans, causing blood to be retained in the body's core to keep internal organs warm, but at the expense of those extremities at the furthest distance from the heart, causing frostbite.

People adapted to live at high altitude in rarified air containing less oxygen.

Vitamin D played a major role in natural adaptation. Humans can't live without vitamin D, and there are but *two sources* of it: sunlight, which stimulates the production of the vitamin by the body, and food. In the absence back then of domesticated cows and goats, the milk of which contains vitamin D, and in those geographic areas in which there were no fish with vitamin D content, and where the sun's rays were far less direct than nearer the Equator, an adaptation had to occur for the sake of survival.

To compensate for the lack of sunlight and appropriate food sources, the skin's melanocytes – which create the brownish pigment melanin and which are present in comparable numbers in *all* skin of *all* the peoples of the Earth – adapted over multiple generations to produce less melanin. This natural adaptation occurred for the sole purpose of allowing more sunlight to penetrate to stimulate the production of vitamin D.

Obviously, those who remained in equatorial climes had no need of such an adaptation to survive, and those but a little further away from the equator needed less adaptation than those closer. **This resulted in a broad range of skin pigmentation among humans.**

The **human eye** also adapted to require less pigmentation the further away from the equator one travels – and thus toward less direct and less damaging sun and ultraviolet light. In fact, *every* human has the *same structural eye color*, which is light and variable between blues and greens and rarely, violet.

No doubt the fact that every human has the same structural color to their eyes would come as a big surprise to most. Those with darker eyes merely have eyes adapted for sunnier climates, protecting them from damaging UV light with additional pigment.

There are even plastic surgeons who can remove the pigment from one's eyes with ostensibly no side effect such that anyone – not that they should wish to do so – could reveal their own bluish or greenish eyes!

This same adaptation is even responsible for **hair color.** When we grow old, our hair loses pigmentation and we all revert to the structural color of our hair keratin: white. If we choose, we can restore the pigment in our hair.

However, these relatively simple adaptations in no way separated man into a different species or race according to the leading scientists.

Those that due to their location breathed cold, dry air (in northern Europe, for example) developed longer nose bridges and narrower nostrils, permitting cold, dry air to be warmed and humidified prior to entry into the bronchial tubes of the lungs.

There are numerous such examples, but the bottom line is that *all natural adaptation and the relatively minor changes in appearance that resulted were the consequence of the isolation of populations exposed to various environmental pressures over time.* However, from the numerous studies undertaken, these natural adaptations most certainly did not create differing species nor did they even give rise to distinct races. We are instead one species without distinction with the exception merely of minor adaptations to climate and environmental pressures.

The mistaken notion of "race" came about merely because of a lack of understanding that a few visible environmental adaptations somehow constituted beings that differed in some substantial, important way, and as such, people could be categorized in this manner. *As it turns out, this was an entirely artificial and arbitrary system not supported by in-depth studies of the totality of the underlying genetic structure and coding.*

Biologists who specialize in categorizing all living things into family, genus and species have determined – and are unified in this assessment – that every living person on Earth is of the exact same family, genus and species: homo sapiens sapiens. In other words, *we are all minor variations on a single theme, and not members of populations who can be grouped into arbitrary, so-called "races."*

You've probably heard the phrase: "There's only one race: the human race." Most people who have repeated this phrase thought of it as a figurative statement or a philosophy, not as the literal truth, but as we now know, it is exactly that: the literal truth.

Unfortunately, over the centuries, in the absence of this knowledge, people focused only on a few visible physical traits, such as skin, for example, and based their notion of separate races solely on these superficial observations.

Biologists now tell us that these very few visible traits are of no more significance – and in many cases, far less so – than any other genetic trait that has been passed on from human to human over the millennia – and most of them are invisible to the naked eye. There are hundreds of genetic markers, which only when taken in their totality, can give an accurate picture of a person's genetic make-up.

The number of combinations are virtually infinite, but at our core, we are all 99.9% identical. There are perhaps 60 novel mutations per human being. Multiply that by 7.5 billion people living today and 450 billion variants are possible. But again variations on a central theme, on a central blueprint.

Again, in the absence of this knowledge, isolated ancient peoples, who noticed another group suddenly appearing in their region whose appearance differed slightly from their own group – rather than correctly identifying the newcomers as kin, as brethren, as family – saw only the visible differences and developed an "us versus them" mentality. The seeds of mistrust were instantly planted. Throughout history, this same fearful reaction to strangers who looked different occurred repeatedly.

Based on this fear, and perhaps due to the inherent desire that gave rise to the adage "birds of a feather flock together," we chose to self-isolate or self-segregate from anyone who appeared different. Living apart, we would naturally tend to develop differing mores, customs, languages, etc. These cultural differences are mostly celebrated today, but have also been instruments of division.

When the *entire* genetic code of individuals is studied, there is a wide range of matching markers that do not at all correspond to the visible, superficial physical traits of mankind. In fact, it has been determined **that the genetic difference between the two people on earth who are the least like each other actually only differ by something like 1/10 of one percent of their respective DNA!**

It has been scientifically proven through DNA research that no human being on earth is any more distantly related to another human being than their 50th cousin...and other research places that least related human to you at no more than your 16th cousin!

In one surprising discovery, there is a tribe in Africa (don't recall where they are located) which has been found to have specific matching genetic markers to people living in Finland, yet these specific traits are not present in neighboring African peoples. Superficially, there are probably no people who resemble each other less, and yet it is clear that the kinship is real!

Genetic studies have also located another African tribe living in Ethiopia which is comprised of the descendants of the small subset of Jewish people who left modern day Israel to escape an invasion of their lands more than several thousand years ago. The genetic marker for carriers of Tay-Sachs disease is one genetic marker of Jewish people, so to find this in any group, regardless of their differing superficial appearance, denotes Jewish kinship.

By doing basic math, one can determine that if every human being has two parents, and each of your parents has two parents each, the number of your ancestors doubles every generation. But this doubling process runs into trouble rather quickly. By the time you would reach say the age of Charlemagne, you would already have reached one trillion people! The inevitable conclusion is that many of these ancestors are at that point the same people found on everyone else's family tree.

Eventually, as we move further back in time, we arrive at a very small group of individuals who are the common ancestors of everyone on earth!

Scientists now know that it is no more valid to attribute certain superficial characteristics as constituting a different race than it is to create separate races for any other observable physical trait.

In keeping with the presumption that *observable differences*, minor though they are, validly constitute a division called "race," we would be obligated to create and recognize countless other races: the left-handed race, the right-handed race, the blue-eyed race, the green-eyed race, the hazel-eyed race, the gray-eyed race, the brown-eyed race, the black-eyed race, the dark hair with blue eyes race, the light-haired, brown-eyed race, the dark-haired, moderately dark skin, but light-eyed race, the quick-to-anger race, the slow-to-anger race, the extroverted race, the introverted race, ad infinitum!

The list is endless...and that is only observable characteristics.

Scientists point out that all traits – both visible and invisible to the naked eye - are instead but *minor variants of an essentially common genetic code possessed by all living humans.*

Another variant on the common genetic human code that lends further credence to this assertion is **blood type**.

For different "races" to actually exist, it certainly stands to reason that members of any given race must share more than a few relatively inconsequential, superficial traits. For peoples to fit neatly into distinct categories, certain core, invisible, but known physiological traits must also be common among them without exception.

For example, one race would possess a certain blood type, and another would contain its own unique type. But this is not the case. Every group on earth shares an identical set of blood types.

A person with the lightest-pigmented skin possible is equally as likely – in a random matching – to be able to accept a blood transfusion from an African in the remotest part of the African continent as he is from his own brother. In fact, were he to limit his search to *only* his own brother, his odds of finding a match would not be as good as by selecting from **any** sizable *group* of individuals in any part of the world, where the probability would be 100%.

Were one to cling to the myth of race, it would be nearly as problematic for that person to explain the existence of *albinism* (the lack of pigmentation), a condition which occurs in both the so-called "Negroid" and "Caucasian" races.

These and numerous other examples make it impossible to defend the notion of "race."

Yet *we not only assume the existence of race, we have legitimized it in many aspects of our society*. Consider in what ways the myth of race continues to be used illogically and even absurdly in our own American society of the 21st century.

We are still required (or at least strongly encouraged) to select from a pre-designated list of "races" on a myriad of forms. If a person happens to be precisely half "Caucasian" and half "Negroid," terms that are now extinct among those whose blinders to reality have been lifted, it's commonly assumed that this person is "African-American." Even if race were a reality, wouldn't it be an equally valid choice to select "Caucasian"? What about if the person is 75% Caucasian? What about 90% Caucasian? Where do we draw the line? Why is it important that a line of any kind exist at all?

My own grandchildren are on one side (i.e. 50% total): 25% Filipino (supposedly "Mongoloid" / Asian), 25% divided equally among Danish, French and Italian ("Caucasian"), and on the other side (50% total) they are equally English / Scottish / Irish and Dutch ("Caucasian"). Their so-called visible "Mongoloid" features, while not pronounced, are certainly noticeable. I guess if we can see them (then again, maybe some people would, and others might not clearly see the distinctions), they're of the "Asian" race then, and should choose accordingly on a census form or to permit the government to ensure that all are being treated fairly in obtaining employment and mortgage financing. Really?

Even were "race" a fact instead of the illusion we now know it to be, my grandchildren are genetically less "Asian" than "Caucasian" - regardless of their appearance.

Allow me to digress into a bit of humor for a moment. I have a great idea that should solve the problem!

People should simply select whichever "race" they themselves most personally identify with.

For example, we could use the cultural markers of one's choice of music and manner of dress. That way, the rapper Eminem can select African-American, and Condoleeza Rice, who is a pianist formally trained in European classical music, and a former stateswoman among predominantly "Caucasian" counterparts can whimsically select "Caucasian / White." [In fact, a recent genetic test performed on her blood determined that her ancestry is 40% European.]

That would clarify everything, wouldn't it? Uh, O.K., I guess not. **Scrap that idea!**

***Question*: So if there is no such thing as "race," why is it that people call each other "racist"?**

The answer to that is that obviously, *if "race" itself was birthed from ignorance, so, too, would any notions associated with it.* Even so, perhaps one could attempt to justify it as a semi-valid term for someone who, though mistaken, genuinely believes in "race," thinks and acts accordingly in a divisive manner.

But *for those of us who clearly see the interconnectedness, the brotherhood / sisterhood, the single family that all the peoples of the world collectively represent, the word "**racist**" becomes an entirely meaningless term.*

To the point, if "race" is an illusion as scientists have proven, and I know and believe that, and consider every human being an extended member of my own family, literally my cousin, I could never be considered a racist. **It's an entirely illogical premise.**

We all recall Alex Hailey's book and the subsequent popular television series called "Roots" back in the 70s. Hailey was able to trace his own roots back to a region and even a specific village in Africa. But if we step back and look at the big picture rather than looking a mere 300 or so years into the recent past, we might more properly re-title the series "**Recent Roots.**"

We now know that every single human on Earth came from Africa, and as such, *every single American citizen is literally African-American!*

I'll bet you didn't see that one coming!

Their paths from Africa to America might differ. They might have come directly from Africa in the 1700s and 1800s, or earlier or later, from Europe, or by the time our country came into existence, they might have already been living here for millennia if they were among the descendants of that small group that crossed the Bering Strait to become a member of a Native American "tribe."

As Martin Luther King's granddaughter Alveda King stated recently: "We may have all come here on different boats, but we're all in the same boat now!"

But regardless of the path taken or physical appearance, **the truth is that 100% of the American population can trace their roots to Africa, their lineage to the same "parents," and we are therefore, factually, all of us, African-Americans, and all of us members of the same extended family.**

Again, scientists have determined that every single human being on Earth is, at most, your 50th cousin! And more likely, a much closer relative than that!

Obviously, *given enough time, the supposed visible traits of every so-called "race" will be so thoroughly mixed that try as they might, those ignorant of the truth concerning the illusion of "race" will have no choice but to forever abandon any attempt at racial classification.*

We will finally be able to toss this unsupported, divisive notion of "race" – born of our own ignorance – into the dust bin of history.

* * *

Hopefully for future reference, because you will likely need this book in your life more than once…to defend your newly found position. Oh yes, you will likely speak about this to others **as the most zealous converts are always the most recent ones.**

Rather than trying to become an Apologist (defender) of your new found perspective and attitude, just ask them to go purchase my book!

Obviously, I would like you to believe in your metamorphosis in thinking enough to recommend my book, to launch it to "viral" status…and in the process make me filthy rich!

In the voice of the millennial: LOL! Couldn't resist!

A Black Brazilian mother poses with her five children.

Her husband is also black. Three of her children are affected by a condition known as Albinism, lacking the genes for melanin production.

Melanin production in varying amounts creates an infinite set of hues and shades in our skin.

Absent any pigmentation showing through our translucent skin from just under the surface, it is white. Our hair absent pigmentation is white. Our eyes, absent the pigmentation that creates a variety of darker hues in our irises, are structurally bluish-green or violet.

Melanin is needed primarily for the protection of our skin, eyes and hair from the damage from ultraviolet light from the Sun that causes cataracts, melanoma (skin cancer), etc. However, we need the Sun's light to stimulate our own body's production of Vitamin D, which is crucial to human health.

As such, with every degree of latitude north or south of the equator, melanin production decreases accordingly. This explains the continuous shading of one member of the human family and his (literal) cousins in other latitudes.

CHAPTER 5

THE CURRENT POLITICS AND PHILOSOPHY OF "RACIAL" DIVISION

"We must learn to live together as brothers {and sisters} or perish together as fools!"
~ Martin Luther King, Jr.

Spring irises bloom.
The **caged bird** no longer sings
The knee on his throat.
~ Kamand Kojouri

As you read Chapter Two, you might have noticed how the colors of the pills seemed to align with the all-too-familiar political colors: **Red States** (Republican) versus **Blue States** (Democrat).

That the *Blue Pill* is held in the "Left" hand. That the **Red Pill** is held in the "**Right**" hand. This convenient political alignment, I believe, was purely coincidental in the context of the movie “The Matrix," but it was entirely intentional that I emphasized it here.

Wait, wait, hold on…there you go again, Christian, I sense you thinking. I thought this was a journey, an opportunity to learn more about the flawed notion of “race.”

Rest assured: No, I haven't pulled a “bait and switch” on you. It is only that *most Americans' perspectives on race are inexorably intertwined with their political views.* We need devote our time together on only one such chapter to put it all into political context, but that's it!

Dare to scratch the surface of these good men and women I highlight here, and you are sure to discover *one common thread*, despite their entirely unique backgrounds and life experiences, and that would be that *they are consciously or subconsciously driven by the Judeo-Christian ethics of love for their fellow man or they came to feel love from whatever source.* But make no mistake about it: You don't feel this passionate about anything unless it comes from fear, hate or **love**.

Knowing most of these good men and women of character from their other writings, speeches and interviews, I can tell you confidently that **it comes from a place of *love* – which is the only power or force in this universe capable of making any lasting, real or positive change in the world!**

I am *not* going to apologize for my own personal perspective on politics or life, my own viewpoint. You can read and accept it…or reject it. Each of us individually has the freedom of personal choice, most especially when it comes to our own thinking.

As opposed to seeking out victimization, these luminaries (a word which has meaning within the context of Chapter Seven as you will soon discover) propose lifting their brethren from "bondage" through a tried-and-proven **value system**.

Shelby Steel recounts how his father had a third-grade education, but through hard work, study, persistent application of whatever knowledge he gleaned, renovating a rack-shamble house and renting it out, taking his positive cash flow and repeating the process several times, survived, made a good living and supported a family.

Family is another recurring value and theme throughout these passages. The family is held up almost as sacrosanct, inviolable as a foundational, societal value worth promoting and preserving.

In contrast, there are many in our society today that seem hell-bent on the destruction of the family unit with policies promoting dependence on welfare, a disincentive to rise up and advance one's status in life economically; easy access to abortion, in effect ensuring that the black population remains in the minority relative to Caucasians, Asians, Hispanics.

Policies / positions / stances such as these create unintended consequences such as fatherless households and entrenchment of the poverty cycle.

The "**One Party Monopoly**" in the control and policy-making for America's big cities must end. It's time to take a different path.

* * *

Video Transcription List of prominent (mostly) "black" / "African-American" voices in America on the false constructs of "race" and "systemic racism":

- "The Myth of Race" - Vox
- "Shelby Steele on goal of protests over George Floyd's death: It's about victimhood and power" - Fox News
- "The Origin of Race in the USA" - Origin of Everything
- "Why we should never apologize for our race" - Good Patriot
- "I do not support George Floyd & here's why" Candace Owens
- David Rubin and Larry Elder on Conservatives, Black Lives Matter, and Racism
- "Morgan Freeman—the role race & genetics play in wealth—CNN"
- "Debunking Systemic Racism & Having Common Decency (Pt. 2) Thomas Sowell"
- "Larry Elder: There is Zero Tolerance for Racism Against Blacks in America [so why do some insist we have "systemic racism" in America?]"
- "Live, Liberty & Levin—Sunday, June 14" - Dr. Carol Swain (segment one) and Dr. Wilfred Reilly (segment two)

* * *

"The Myth of Race; "Race" debunked in 3 minutes," Vox https://youtu.be/VnfKgffCZ7U

You may think you know exactly what race you are, but how would you prove it if somebody disagreed with you?

The fact is, even though race drives a lot of social and political outcomes, race *isn't real.*

One of the first people to attempt to categorize humans according to race was a German scientist, around 1776. He came up with five different groups according to physical appearance and geographic origin of their ancestors. Americans of European descent eagerly bought into this type of thinking around the same time.

Some historians have [even] said the idea that there are different "races" helped them resolve the contradiction between a natural right to freedom and the fact of slavery.

If "whites" were their own distinct category, then they could feel a lot better about denying freedom to people who they labeled "black" and decided were fundamentally different.

But, as political priorities change, definitions of race in America adjust right along with them.

For example, if you were of Mexican birth or ancestry in the United States in 1929, you were considered white. Then, the 1930 census changed that to "non-white" to limit immigration. Later, when the US needed to increase its labor force during World War II, these people were switched back to white. And what it took to be "black" once varied so widely throughout the country, from one quarter, to one sixteenth, to the infamous "one drop [of blood]" of African ancestry, that people could actually change races just by crossing state lines!

Then, suddenly, in 2000, the government decided that Americans could be more than one race and added a 'multiracial' category to the census. This has left many Americans scratching their heads when it comes to selecting who they are. As many as 6.2% of census respondents selected "some other race" in the 2010 survey.

The idea that someone might look one way, and identify another way, or that they might be really hard to place in a racial category, is not new. This is why there was a public debate about whether MSNBC's Karen Finney could say she was black, or how we can't even agree on the racial label assigned to the President of the United States!

Of course, many people feel their racial identity is very clear and very permanent, but the fact that some people have changed theirs, and that nobody can really argue with them, shows how shaky the very idea of "race" is.

THIS IS ALL BECAUSE THERE ISN'T A RACE CHROMOSOME IN OUR DNA THAT PEOPLE CAN POINT TO. IT SIMPLY DOESN'T EXIST!

When the medical community links race to health outcomes, it's really just using race as a substitute for other factors, such as where your ancestors came from, or the experiences of people who may have been put in the same racial group as you. Dorothy Roberts explains that *sickle-cell anemia* is a prime example of this.

The disease is linked to areas with high rates of malaria, which include some parts of Europe and Asia, in addition to Africa. *It's not actually about race at all.*

This of course does *not* mean that the [false construct]

concept of "race" isn't hugely important in our lives. *The racial categories to which [we choose or are] assigned can determine real life experiences. They can drive political outcomes, and they can even make the difference between life and death.*

But understanding that racial categories are made up can give us an important perspective on where racism came from in the first place.

* * *

Chronicle / Eric Lu

"Shelby Steele on the goal of protests over George Floyd's death: It's about victimhood and power," Fox News

Mark Levin, Host; Shelby Steele, Senior Fellow, Hoover Institution, Guest https://www.foxnews.com/transcript/shelby-steele-on-goal-of-protests-over-george-floyds-death-its-about-victimhood-and-power

Author: Shelby Steele is a "black" Conservative who contends that blacks have been "twice betrayed": first by slavery and oppression and then by group preferences mandated by the government (e.g. affirmative action), which discourage self-agency and personal responsibility in "blacks."

(Start at 03:55)

Steele: And so when people start to talk about "**systemic racism**," built into the system, what they're really doing is *expanding the territory of entitlement*. We want more. We want more. We want the society to give us more, to help us, and so forth. Society is responsible for us, because racism is so systemic.

Well, that's a corruption! And I know it's a corruption because the truth of the matter is, blacks have never been less oppressed than they are today.

Opportunity is around every corner, and in all of this no one ever stops to say, "Well, you're unhappy about where minorities are at in American life and blacks continue to be at the bottom of most socioeconomic measures?" You're unhappy about that. Well, why don't you take some responsibility for it? Why don't you take more responsibility?

I would be happy to look at all the usual bad guys, the police and so forth, if we had the nerve, the courage to look at black people [as well]. To look at black Americans, minority Americans, and say, "you're not carrying your own weight! You are going to go have a fit, and a tantrum, and demonstrate, and so forth.

[**Value system**]

And yet, are you teaching your child to read?

Are you making sure that the school down the street actually educates your child?

Are you becoming educated and following a dream in life and making things happen for yourself?

Or are you saying, "I'm a victim, and I'm old, and the entitlement is inadequate, and I need to be given more; and after all, you know, you whites, that racism has been here for four centuries, and slavery, and so forth, and so it's time for you to give [more] to me."

Well, that's an exhausted, fruitless, empty strategy to take!

We've been on that path since the '60s and we are farther behind than we've ever been and we keep blaming it on racism and blaming it on the police. I'm exhausted with that.

I grew up in a time when there was real segregation. And blacks during the 50s and so forth took a lot of responsibility for their lives because the government didn't.

My father bought three ramshackle houses, rebuilt them, rented them out, kept clawing his way up the ladder. A man with a third grade education from the South.

Well, what Civil Rights Bill is going to replace that? That value system?

And he was not exceptional. Across the community we lived in, those were the values and that is the problem: we have allowed ourselves to be enabled in avoiding our real problems by a guilty, white society that keeps using us, exploiting us as victims, to say that it's wonderful and it's overcoming racism and so forth.

If you really care about what how minorities do, why don't you ask them to do it? Why don't you ask them to drop the pretense?

There's always going to be some racism in every society. My own sense is that it's endemic to the human condition. We will always have to watch out for it. As I like to say, stupidity is also endemic to the human condition and we have to watch out for that, too. That is no excuse for us being where we are right now in American life. We have let this sort of guilty society and our grievance industry put us in this impossible position where we are a permanent underclass. Before the 60s, there was no black underclass. That's a new phenomenon. Public housing completely supported by the government.

(Stop at 9:03)

* * *

"The Origin of Race in the USA," Origin of Everything
https://youtu.be/CVxAlmAPHec?t=249

(Start at 04:09)

In order to continue to justify slavery, we start to see the pseudo-science of "race" emerge that connected physical features, behavior, and legal rights, right around the 18th century when colonial use of slaves was expanding.

Anthropologist Audrey Smedley notes that "scientific"

ideas about physical appearance and racial difference in the 18th century were largely "folk" ideas used to justify already existent social norms.

So as a result of a desire to perpetuate systems of exploitation, more and more distinctions were made about the supposed differences amongst races, primarily the differences of black people from their white counterparts. This evolution of race became more concretized after social structures of slavery were in place, and not before, and was solidified by the Enlightenment.

Which brings us to our third question: How did the Enlightenment impact definitions of race?

The Enlightenment was a period of primarily European thought and ideological development that saw the emergence of some key concepts that tie back into today's discussion.

First: there was a push in scientific communities to categorize the natural world using "reason" and creating elaborate hierarchical systems that emphasized the similarities between different species and subgroups and the inherent differences amongst others, and race was fitted into this same mold. As European theorists looked to classify the world into "scientific" groupings, physical markers that were already established social norms through enslavement and genocide were ways that they sought to "prove" that this was the "natural" order and not a "social construction."

For example, Thomas Jefferson, who was a proponent of concepts like individual liberty and freedom for white men, or those he considered his peers, also made claims that black slaves required less sleep than their European counterparts to justify excruciatingly long and inhumane work hours.

And Samuel Cartwright, who falsely claimed that "drapetomania" was a mental defect that caused enslaved black people to run away from slavery, as if wanting to escape a lifetime of enslavement was...illogical?

The Enlightenment formulation of history also played a crucial role in the development of social ideologies of race.

Kant, Hegel, and other philosophers of their day claimed that certain racial groups stood outside of history, or had no history, and this included all groups that they considered non-white or outside of European ideals of modernity. This meant that groups that were devoid of history and culture were inherently less valuable, and therefore subordinate to other races. They were cast as the natural sacrifices of "progress."

These assumptions were also codified into law in the 18th and 19th century.

The first naturalization laws of the United States in 1790, limited naturalized citizenship to "free white persons" and excluded other groups. Children born of enslaved mothers were said to inherit the legal statuses of their mother, effectively keeping them in bondage perpetually. And Native Americans were often denied legal property rights, which helped to expedite the process of Westward expansion across the North American continent. And "anti-miscegenation" laws were drafted in order to assure that people from different racial backgrounds did not intermarry or have children in order to protect ideals of racial purity.

But these racial categorizations didn't always neatly align with skin tone. In his book Whiteness of a Different Color: European Immigrants and the Alchemy of Race, historian Matthew Jacobson notes that, in the U.S., "White" or "Caucasian" was not always considered a unified race composed of anyone of European descent. Whiteness was often considered exclusive to Anglo-Saxon descendants, while other European groups were broken into different ethnic categories such as "Celt," "Slavs," "Iberics," and "Hebrews," which were considered separate races from the 1840s to the early 20th century.

But, in the 1920s, when there was a stemming of migration from Europe, these different races were subsumed into one category called "whiteness" to shore up a cultural majority against other racial groups and immigrants. And this persisted throughout the 20th century.

So how does it all add up?

Well, **race started as a marker of kinship. But then we see it shift to become less about familial inheritance and more focused on physical indicators due to the rise of Enlightenment reasoning and labor exploitation.**

But where does that leave us today?

In the past several decades, scholars have noted both an uptick in identifying with your ethnic history, for example pride in your country or culture of origin, instead of just being "white" or "black." But we've also seen a revival of outdated theories of race being biologically determined and attached to certain traits, like considering all people of one physical type as prone to certain behaviors. But regardless of how people identify themselves, "race" continues to be a complex topic of discussion and debate.

(Stop at 08:59)

* * *

"Why we should never apologize for our race," Good Patriot
https://youtu.be/0wrzoIBC7KY

Hi America, I'm "Good Patriot"…and I'm a racist piece of trash. Well, not a regular racist, I'm actually a "subconscious racist." I'm so racist that I don't even know how racist I am, or at least that's what I've been told.

So, to destroy racism, which apparently only exists against black people, I'm gonna have to do a few things.

Number one: Put a black box on my social media pages. This will ensure that I start working off my penance for being born white, and it will also make sure that my friends that aren't racists know that even though I've never done anything racist that they know, that I'm not racist either, and then somehow that's gonna help the black people.

Number two: As a white person I need to validate the existence of black people by saying the magic words "black lives matter." Saying that *actually* shows that I think that my white opinion is superior and is needed to validate a black person's existence, and worse, and black people acknowledging that they need to hear it from me is actually saying that they find their worth in my white opinion, but let's just ignore all the mental gymnastics of that and just say it anyway. Black. Lives. Matter!

There you go, racism has been annihilated!

Okay let's cut the crap and get down to business!

I haven't heard one person say that George Floyd's death wasn't unjust and tragic and I have not seen any evidence that there's a racial motivation behind his death, but the media told you to feel bad, to get mad, and that it was okay to do stupid things because you've had years of anger and oppression built up. They lied and you believed it. Why haven't you been mad at the dozen or two blacks killed in Chicago for the 57th weekend in a row? In 2018, white people committed violent crimes against black people almost 60,000 times. In the same year, black people committed violent crimes against black people almost 400,000 times, but y'all just want to talk about [white on black crime], not [black on black crime].

I guess black lives don't *really* matter unless they're injured or killed by whites and then the media tells you to be mad about it.

Are you recognizing a pattern here? The media owns you. They tell you how to think and how to feel. This is a tragic death used by the higher ups to bring destruction and division and you are falling right into it. If you really cared about brown and black lives you would bring down Planned Parenthood. They were literally built to murder black lives, but you go on and righteously boast your bull crap hashtags thinking that you're doing good and at the same time you're voting for more abortion rights so that they can rip apart black babies.

Have you ever heard of the Barbary Slave Trade?

That's the one where, for hundreds of years, millions of white Europeans were kidnapped and enslaved in North Africa! Kind of weird that we never hear talk about that example of slavery, right?

How about the police killings of Kelly Thomas and Duncan Lemp? Haven't heard of them? I'm not surprised. Did you know that black on white crime is over nine times higher than white on black crime? I bet you didn't. That's because it doesn't fit their agenda to make you think that black and brown people are the perpetrators and that white people are privileged, power-hungry murderers.

Did you also know that the CIA has been conducting psychological operations on us for decades using our media?

This isn't about race. This is about distracting you from the real enemy and manipulating you to think in a certain way.

To the whites: Who of you have actually acted out in an evil way against a black person just because they're black? I haven't, ever, and I'm guessing that a very large majority of you haven't either.

You're not racist, so stop feeling guilty.

I don't have to prove my non-racism to anybody; I just live it. I don't owe black people, I don't owe the media, I don't owe some stupid social media influencer's celebrity campaign *anything* and I will not ever apologize simply for being white, and I will never apologize that my hard work got me to where I am.

These looters and rioters have sold their integrity for $15 an hour and a stolen pair of shoes. They are not oppressed. White, black, or brown: they are thieves, murderers, and they deserve no sympathy. I have gotten on my knees and wept over these people and prayed that God removes the blindness from their eyes and that they turn from their evil ways, but so help me, God, if one of you comes near me or someone near me, and threatens our livelihood or our lives, I will not hesitate to neutralize your threat.

To the black people that got their peaceful protest hijacked by these evil animals, I am truly sorry! George's death was a disgusting crime, but there are also disgusting crimes committed every day against all races. We just don't get told to be upset about them.

Let this wake you up to the fact that you were being used as a pawn in a very calculated game of good versus evil. We all are. And **that's why *we need to stick together and stop letting them dictate how we feel, what we do, and how we think.*** Don't believe the lie that you can blame your way to success and riot your way to justice.

I'm tired of all this bull crap and we need to stop being angry at each other and be angry at the people that are actually systematically trying to oppress and control us, and until we do that, they're gonna win.

To the believers out there, put on the full armor of God and get on your face in prayer and pray over our country and our world because we are in a serious battle right now.

I hope you realize that I'm coming from a place of tough love because that's what's needed right now. I'm done with political correctness and sugarcoating!

"I do not support George Floyd & here's why," Candace Owens
https://youtu.be/JtPfoEvNJ74?t=540

(Start at 09:00)

I want to be clear: this is not a defense for Derek Chauvin. No one that I have spoken, to no one in the news, is defending Derek Chauvin. He is getting what he has coming to him.

But why are we pretending that this criminal should be upheld as a martyr in Black America? A martyr for a fake narrative, by the way.

Police brutality, *racially motivated police brutality, is a myth.*

Not only are we using this death and allowing it to cause these riots and protests pretending this was some upstanding citizen the black community who was tackled down and killed for no reason. Not only are we allowing it to inspire riots. Riots in which black people are dying, in which actual upstanding black citizens are dying.

I'm sure you guys have all now seen David [Dorn], the ["black" police captain] that just was shot and killed because he was trying to protect a pawn shop from looters. An upstanding citizen, an actual head of a police [department]. 77 years old, did everything right.

(10:16)

So we now have to kill upstanding black citizens because a non-upstanding black citizen, a career black criminal, died. Did he deserve to die in that manner? No. I can't say it enough. No, he didn't deserve to die in that manner, but I will be damned if the rest of us upstanding black citizens have to suffer because of this incident that rarely ever happens in America.

So here are some numbers for you people that are still believing that real, racially-motivated police brutality is a real thing.

First and foremost, you have a 25% higher chance as a violent white criminal of dying at the hands of a police officer than you do as a black criminal. Last year, a total of 9 unarmed black men were killed by police officers and 19

white men were killed by police officers. For those of you that aren't good at mathematics you might be thinking, "Oh, but Candace, white people represent 60% of the population and black people represent just 13% of the population."

(11:21)

It doesn't matter what percentage of the population you represent. It matters what percentage of the violent criminal community you represent, and unfortunately the black community commits a disproportionate amount of crimes compared to the white community. 6% of the population, black men, 6% of the population account for 44% of all murders in this country according to 2018 statistics.

That is what you call a gap.

And yet white people, who represent 60% of the population, we represent 13%, black men are 6%, represent 50% of all the murders. That makes no sense! A six-point variation in a community where we are extreme minorities, we commit 50% of all violent offenses, evenly split and we're only 13% of the population! So [therefore] we have a lot more encounters with police officers.

And don't say that police officers are coming around because we're black. I'm talking about violent criminals. I'm talking about murder, 44% of murders. It's real statistics. The police officers have way more to be fearful of in the black community than the other way around. On average, a police officer is 18 and a half times more likely to be killed by a black person than the other way around!

(12:38)

This entire narrative is complete smoke and mirrors. It's all made up. It's just election fodder. It's "white" versus "black" because it's an election year, not because black Americans are suffering at the hands of police officers more than white Americans.

Do some police officers do the wrong thing? *Yes.* I don't think there's anybody in the world who has not encountered a police officer and thought, "this person is an absolute jerk who is power tripping," whether you are black or white. We know they exist and we know they're always going to exist, by the way, because they're human beings and sometimes human beings suck.

In fact, if you want to attack a community for accidental slayings or brutality, did you know that doctors "accidentally kill" a quarter of a million people every year because of mistakes? Do you know that there are doctors that have been arrested for being serial killers? Just for killing people because they wanted to. Do we protest and boycott doctors? Do we assume all doctors or horrible human beings because some doctors are? Or do we realize that society is not perfectible?

(13:41)

People suck in every profession. It is no excuse to paint society with a broad brush. **It is certainly no excuse to accept a Democrat narrative that black people are being disproportionately hunted down by police officers because of the color of their skin.**

You want to know the best way to avoid being brutalized by a police officer? It's to limit the amount of encounters you need to have with them, especially when it comes to violent crimes. I am not going to stand for this continual bottom-feeding narrative of us martyring people that have had five, six, seven stints in prison, and then pretending they were upstanding heroes to our community. It's bullshit. Excuse my language, it's absolute bullshit, and I'm tired of it. I'm tired of having to play pretend. I'm tired of sitting here and being called a coon or an Uncle Tom because I won't play this bottom-feeding game with you.

If you want to hang up posters of criminals on your wall and talk about them as your martyrs, do it. Do it! You can do it.

Kobe Bryant was my idol. I'll keep a photo of Kobe Bryant, you can keep one of George Floyd and pretend he was an upstanding human being that just once or twice put a gun across a pregnant woman's belly. Could you imagine that woman right now, that black woman by the way? Watching everybody pretend this person was an upstanding human being, who just at the age of 42 and [after] five prison stints was going to get his life together. I mean it's embarrassing. We are embarrassing in that regard.

This is why we have a cycle and a toxic culture, because nobody wants to tell the truth in black America. It's so easy to be a victim. It's so easy to ask white people to bow down, and apologize, and do all these things for us. It's crap! It's a lie; it's a farce.

(15:16)

Our biggest problem is us.

It's why we don't talk about it when "black-on-black" crime happens. It's why we don't talk about it when 40 black people are killed in one weekend during Memorial weekend in Chicago. We don't talk about any of that stuff. We don't want talk about Baltimore. We don't talk about New Jersey. We don't want to talk about any of these places where black people are being slaughtered by other blacks, because that would mean that we had to be personally accountable.

That would mean personal responsibility and we don't do personal responsibility in our community. We don't do it. We blame white people [instead]. We only point a camera to white people when they do something. Even though we [commit crimes] at a way higher rate to ourselves. We celebrate our drug dealers. We're the only community that would ever create hashtags to free people from prison because they committed crimes. Like "free Meek Mill," "free this rapper," "free that rapper."

(16:02)

How hard is it to not spend multiple times in prison? How difficult is that? Is that too hard for us? I mean is that way too high of a mountain for us to scale, to do the right thing, to be upstanding citizens?

That is the call to action that I have for Black America with **BLEXIT**.

Why do we keep fulfilling this narrative? What do you

think the perception of us is on the outside? You ever look at the comments, you ever go into like an anonymous blog and see what people say? "Oh, just black people being black people." I see those racist comments. "Oh, just black people gotta riot. Black people gotta be black people. You know how they are. Oh, just black people being ignorant."

That is the perception when people get to be anonymous and talk about us. That's what they think about us. They think that we are the kind of people that will forever uphold criminals as the martyrs of our society. That we will never take account for the things that we do wrong. That we don't have it within us to educate ourselves to get ahead, and that for those of us that actually do it we get called coons.

You got Condoleezza Rice, she's a coon. Larry Elder? He's a coon. Dr. Ben Carson? Brain surgeon. First ever to perform the surgery of splitting twins that are connected at the head. He's a coon. Right? What a loser he is, what a stupid guy he is. Kanye West says he's not going to be told what to do because of the color of his skin. He's a coon, he's lost, he's in "the sunken place."

(17:23)

The "sunken place," that's where we all are because we demand more and we will get more out of this society because we were gonna get ahead. Black Conservatives get ahead because we don't subscribe to this narrative. Because you're not gonna "catch me outside" trying to grab a TV pretending that it's because a "martyr" named George Floyd got killed.

I'm a big believer that no matter what color you are,

you do stupid things, you win stupid prizes. We have to do better. We have to teach our kids better, or we're not gonna get ahead.

(Stop at 17:59)

* * *

David Rubin and Larry Elder on "Conservatives, Black Lives Matter, and Racism"
https://youtu.be/IFqVNPwsLNo?t=1456 | Start at 24:16

Elder: **Tell me what you think the most "systemic racist" issue is.** What is it?

Rubin: Well, I would say that because black people in most cases, in many cases, were descendants of slaves that racism as an institution, that a certain amount of it just exists.

Elder: In 2015? Then give me the most blatant racist example you can come up with right now.

Rubin: I think you could probably find evidence that, in

general, that cops are more willing to shoot if the perpetrator is black than white.

Elder: What's your data? What's your basis for saying that?

Rubin: I know a lot of people would say, "look what's going on in Chicago."

Elder: I know what they would say. I'm talking about what the facts are.

965 people were shot by cops last year and killed. Four percent of them were white cops shooting unarmed blacks. In Chicago in 2011, 21 people were shot and killed by cops. In 2015 there were seven. In Chicago, which is a third black, a third white, and third Hispanic, 70% of the homicides are black on black. [That's] about 40 per month, almost 500 last year in Chicago, and 75% of them are unsolved.

Where is "Black Lives Matter" on that?

The idea that a racist white cop shooting unarmed black people is a peril to black people is BS. It's complete and total BS and the reason for these so-called activists saying this is the assumption that racism remains a major problem in America; the media, CNN, *especially* MSNBC, runs down whenever a black cop shoots somebody and it's some march on Washington. It's ridiculous.

Half the homicides in this country are committed by and against black people. Last year, there were 14,000 homicides (not suicides, I'm talking about homicides), half of them were black, 96% of them black-on-black of that 7,000.

Where's the "Black Lives Matter" people on that?

Rubin: So that there's where you would say that this is purely because of "social justice?" Purely because they want ultimately for people to be angry enough to just keep voting Democrat?

Elder: That's right.

Where's the evidence of a lack of "social justice?"

When a black suspect is killed by a cop, believe me the media is on it, people are watching it, and justice will for the most part occur. In Baltimore, where Freddie Gray was killed. Freddie Gray died in a van; I shouldn't say "was killed." You have a city that's 45% black. City Council is 100% Democrat. The majority of City Council is black. The top cop at the time was black. The number two cop was black. The majority of the command staff was black. The mayor is black. The AG is black. And yet here we are talking about racism. I mean it's absurd.

There was just a study by the University of Washington and it turns out cops were more reluctant, more hesitant to pull the trigger against the black suspect than a white suspect. Probably because of the fear of being accused of racially profiling, and the fear that the Civil Rights establishment was going to come down on him. So, if anything, whites are more likely to be shot by a cop under certain circumstances than is a black person. In the last 30 or 40 years, the percentage of suspects who are black, killed by cops, has declined 75%. However, the percentage of whites killed by cops has flat-lined. If anything, people

are more concerned about shooting black people for fear that they're going to be called racist.

Almost every one of these incidents, whether it's Eric Gardner in New York, who died because he was selling "loosies" and resisted arrest, whether it's Tamir Rice in Cleveland who was twirling around the gun, whether it's Michael Brown in Ferguson who had just committed a strong-armed robbery, **almost every one of these incidents involves somebody resisting arrest.**

Why don't you just do what the police tell you? My dad said, when I get pulled over, have my hand at ten o'clock, have my hand at two o'clock, say "yes sir," say "no sir," make sure my paperwork is in order, and if I feel the cop is mistreating me, get a badge number and deal with it later. If Jesse Jackson and Al Sharpton and Obama and the whole group of them told black people to do [just] that, we'd have a lot fewer of these things to deal with in the first place.

Rubin: I'm with ya. I'm hearing a lot of what you're saying. So as a black Conservative then, you've laid out your case.

Elder: But you haven't laid out yours. I asked you to name the most important example of racism and you gave white cops going after black people. I then gave you the facts for that, so that's nonsense. You must have something else. What else is it? If you think racism remains a problem in America, give it to me.

Rubin: Well I think it remains a problem. **It may not be "systemic,"** in that it's not like you're not being hired because you're black. There's no "systemic" reason, no legal reason that exists, that kind of thing. I think that racism as a general theory exists.

Elder: I need some specifics. You gave me the white cop thing. What else? Give me another example.

Rubin: Well as a black Conservative, tell me how do you get people to come around to…

Elder: No, you're the one who made the assertion that: Yeah, racism remains a major problem in America. I asked you to give me an example and you gave me white cops going after blacks. As far as I'm concerned, you didn't hold it up very well. What's the other argument you have? What's the other thing?

Rubin: Well I don't know that it's "systemic" in that in the sort of macro set.

Elder: I'm not mad, I just want to know what it is you're talking about. So, what? Blacks are not getting into school? BS. We have affirmative action, so a black person with it with an SAT and a GPA of X will get into a school faster and easier than the right person with an SAT or a GPA of X. If going to school there's a route to the middle class you can make an argument that blacks have an easier route to the middle class. If you're talking poverty, the poorer you are the more accessible loans and grants are for you.

The biggest burden that black people have, in my opinion again, is the percentage of blacks, 75% of them, that are *raised without fathers*. That has every other negative social consequence connected to it.

Crime, not being able to compete economically in the country, being more likely to be arrested—that's the number one problem facing the black community. When I hear people tell me about "systemic racism" or "unconscious racism," I always say "give me an example," and almost nobody can do it.

Rubin: So the *family* stuff, I'll follow your logic on that. What can actually be done about that?

Elder: Reverse to welfare state. In 1890, 1900, you look at census reports, a black kid, believe it or not, was slightly *more* likely to be born to a nuclear, intact family than a white kid. Even during slavery, a black kid was more likely to be born under a roof with his biological mother and biological father than *today*.

What's happened is we launched this so-called "War on Poverty" in the 60s, where literally Lyndon Johnson sent people knocking on doors (and I lived in the 60s and people knocked on doors) apprising women of the availability of welfare, provided there was no man in the house. We went from 25% of blacks being born outside of wedlock in '65, to 75% right now.

You look at how much money that we spent on welfare, and the lines are parallel. It was a neutron bomb dropped on this country, not just in the black community but on people in general. At one time only about 5% of whites were born outside of wedlock, now 25% of whites are born outside of wedlock.

I was in college in 1970 and there was a report called the "Moynihan Report: The Negro Family, A Case for National Action," written by a liberal, a man who became a Democratic

Senator from New York. At the time, 25% of black kids were born outside of wedlock. He said, "my God. this number is horrific! If we don't do something about it, it could get even higher." Well, fast forward—25% of white kids are now born outside of wedlock. It is the number one problem in this country and what we've done, in my opinion, is we economically incentivized women to marry the government, we've allowed men to abandon their financial and moral responsibility, and now we have this.

(STOP at 32:44)

* * *

"Morgan Freeman—the role race & genetics play in wealth—CNN"

Don Lemon, Host; Morgan Freeman, Guest
https://youtu.be/wIKGbur0QLU?t=81

(Start at 0:01:21)

Lemon: **Do you think poverty is part of genetics?**

Freeman: No. You can't say that. Personally, **I don't think wealth and genetics have anything to do with each other, actually.** If you're born in the US, it really doesn't matter the condition of your birth. What matters is what you inherit from your nurturing, from your environment. This is just from the standpoint of having been born with little.

Lemon: But environment affects.

Freeman: Say you're born as one of the untouchables in India. There you are. And if you're born into the super-rich? There you are. Here, you can be born into the super-rich, but 3, 4 generations down the line, it's gone.

Lemon: Here's a stat from your show. It says: "the richest 85 people on this planet have as much money as the poorest 3.5 billion."

Freeman: On the planet, however. It isn't far off in America, but we still here don't abandon the idea that it isn't finite. It isn't finite that you're born...where were you born?

Lemon: Baton Rouge, Louisiana. Baton Rouge General Hospital.

Freeman: I was born in Memphis, Tennessee. I had a long haul from where I came from to here, but here we are. The proof of the pudding is in the eating, and here we sit at the dining table.

Lemon: But it's hard, too, when you say that people... because they say, "oh here you go with the 'pull yourself up by the bootstraps thing, you're just being respectable and not everybody can do that.'"

Freeman: Bull-s*. Everybody can. Courage is the key to life itself. There are a lot of people who are born in situations and say "I'll never get out of this," and they won't. I say to people who say, "well I would have liked to have done so, and so, and so," that you could have done it. They say "well I couldn't get out of here." So I say, "man, the bus runs every day!" If you think of it, you can do it. That's the human condition. If we can imagine it, we can do it. I think that's the condition of each individual human.**

Lemon: Can I go back, because I thought it was fascinating when you called it "bull" that people can't pull themselves up. Do you think that race plays a part in wealth distribution, or either a mindset that you can't...

Freeman: Today?

Lemon: Yeah.

Freeman: No. No, I don't. Look at you and me. Why would race have anything to do with it? Put your mind to what you want to do and go for that. It's kind of like religion, to me. It's a good excuse for not getting there.

Lemon: I'll probably get in trouble, but I said to some of my colleagues recently that I know that it's an issue, but **it seems like every single day on television I'm talking about race...and it's because of the news cycle, it's in the news. But sometimes I get so tired of talking about it, I wanna just go, "This is over. Can we just move on?"**

Freeman: And if you talk about it, it exists. It's not like it exists and we refuse to talk about it, but making it a bigger issue than it needs to be is problematic.

(STOP at 6:27)

* * *

"Debunking Systemic Racism & Having Common Decency (Pt. 2) Thomas Sowell"
David Rubin, Host; Thomas Sowell, Economist, Guest
https://youtu.be/5daNMXer2UQ?t=314

(Start at 00:05:14)

Rubin: Are there any laws that are in place right now that you believe are discriminatory one way or another tort towards any community or against any community?

Sowell: I would have to write a much larger book to cover them all. The minimum wage law is absolutely devastating. The policy of saying that you cannot have more kids from one ethnic group disciplined in the school than from another is nonsense. I mean, groups are different from each other and umpteen ways and then to have the presumption that they are the same except for the way their they're treated is nonsense. That's never been true and I don't know why we would think it's true here today.

Rubin: So I do sense that some of what you just said there is bubbling up into the national consciousness because I get a ton of email from black conservatives now, people that feel that they haven't been represented fairly, or that the so-called "leaders" of the black community that are on television all the time are actually preaching the complete reverse of everything that you've said here. Do you sense that there is some sort of growing conservative movement in the black community?

Sowell: Well, there was a time when that that community would consisted of me and Walter Williams. I know Walter used to say "we're the two! We should never fly on the same plane otherwise the whole movement will disappear if the plane goes down!"

Rubin: I mentioned to you before we started that Larry Elder caused my awakening because I was a progressive and I said something to him about systemic racism on-air and he beat me senseless with facts and I had to go back and reassess what was wrong with my thinking.

Sowell: Well you know in one of the in one of the chapters I have a little section about the era of apartheid in South Africa and I had that in there because there's so much argument of "how much racism is there?" and so forth. I said, "let's test this hypothesis in a setting where there's absolutely no doubt of it, and that's apartheid in South Africa. With a government where blacks are not allowed to vote and so forth. And you then apply the economic principles and you find that the economic principles apply in South Africa. Blacks were not allowed by law to be in some occupations more than some percentage, and in some occupations couldn't be hired at all. Some of those occupations where it was illegal to hire them at all there were more blacks hired than there were whites, because **there are economic factors that come in, and you don't just pass a law and that automatically produces the results you want.**

It's the competitiveness of the industry. In a competitive industry, discrimination in the sense that we that we use for anti-discrimination laws, costs the discriminator as well as the others. For example, minimum wage laws. When you have a minimum wage law, you have more people applying for jobs in those categories then there are jobs available because raising the wage rate causes more people to apply and employers to hire fewer because they're more expensive and so we have a chronic surplus. Now if you've got a chronic surplus in an industry, it costs nothing to discriminate, but if you have a competitive market then of course it does cost something. For every person that you discriminate against that was qualified, you've got to hire somebody else and you've gotta raise that pay raise in order to get people in. So I show how **competitive industries have much less discrimination than, say, regulated public utilities.**

Rubin: I was wondering when I was reading it if you were ever going to talk about how now technology is also changing this. We see a lot of these movements for $15 minimum wage and I know why you don't think that's a great idea, but even now we see McDonald's and some of these other places just replacing people with computers.

Sowell: When I grew up in Harlem and when you went into the little neighborhood movie theater there would be a kid who would walk right down the aisle with you with a flashlight to show you to your seat. And now that we have so many compassionate people who want people to be paid a living wage, you actually stumble down the aisle to your seat the best way you can because they're not going to pay that kind of money that's unrelated to productivity.

(Stop at 00:09:55)

* * *

"Larry Elder: There is Zero Tolerance for Racism Against Blacks in America" [so why do we think "systemic racism" is real?!]

https://youtu.be/vrOVYsNeQEs

Regarding these matters that have happened—the man

in Minneapolis who died, was killed by the police. This woman in Central Park who complained about the "African-American gentleman." She works for an investment company called [Franklin] Templeton; they fired her after this video went viral. Four officers fired—not one day of testimony, no grand jury, no DOJ investigation, no temporarily putting them on leave while there's an investigation. They summarily fired. And, by the way, don't think that when they hire their defense lawyers (probably already have) that they're not going to countersue claiming that "we were summarily fired without due process!" I've seen that happen before and I've seen cities that have done that get dinged.

Now, I'm not at all saying that what happened in Minneapolis should have happened. Not at all condoning the actions of these officers. It certainly seems excessive to me! I mean my God, having this man pressed like that with your knee on his neck for four five/seven minutes. And yes he was saying "I can't breathe," and yes, suspects often say "the handcuffs are such, I can't I can't feel my hands," "I can't breathe." People lie all the time, but seven minutes? Five to seven minutes? It seems excessive to me. The point is this (despite what the people at "MSNB-He Haw" say and CNN say) there is ZERO tolerance in this country for racism against black people. Zero.

Former editor of the *National Review* John O'Sullivan once put it this way: "white racism exists, but its social power is weak. The social power against it? Overwhelming." And I'm sure you can think of a lot of incidents, without working up much of a sweat, where that's happened. Remember that relief pitcher John Rocker for the Atlanta Braves who made those asinine comments that were printed in *Sports Illustrated*? Gets canned. Goes from team to team; never did regain his form. Got fired by his team. You remember Al Campanis, the former general manager the Dodgers? On an anniversary of Jackie Robinson becoming a major league player, he was asked on *Nightline*, "why is it there aren't more blacks in positions of management, of responsibility?" He goes "uh, uh, I dunno...." Fired! Done! Michael Richards, the comedian, the one that was on *Seinfeld*, very popular. Doing a stand-up here in LA, gets heckled by some blacks, drops a bunch of "n-words." This man cannot get a table at Fat Burger anymore! Done! I've seen this over and over and over. Paula Deen—done. Made that crack about blacks in that lawsuit. Duane "Dog" Chapman lost his show after making some crack. Zero tolerance. My point is, look at the reaction.

And you know who else believes me, believe what I just now said (although he would not want to admit it)? Eric Holder, the former AG under Barack Obama. He gave a speech in which he referred to what happened to the former owner of the LA Clippers, Donald Sterling, who's on tape at his home, talking to his girlfriend making a bunch of disparaging comments about Magic Johnson and about blacks. Loses his team. And Eric Holder said "you know, when you see *blatant* instances of racism like the Donald Sterling stuff, we got that. We can deal with that. It's the '*pernicious*' racism," he said, "that's the problem today."

He gave three examples, and I've told you about the examples in the past; I won't belabor them. But they were the push for voter ID. Oops, majority of blacks want voter ID. He claimed that because black boys were kicked out disproportionately at schools compared to white boys that was an example of racism. Oops, turns out that irrespective of the racial composition of the school board, irrespective of the race of the teacher, irrespective of the race of the principal black boys are kicked out disproportionately compared to white boys. Nothing to do with race. Unless of course you're prepared to say all these teachers are racist. The third thing he pointed out is that black men who commit a crime compared to white men who commit the same crime will get a longer sentence and that is true. What Eric Holder didn't tell you is that the US Sentencing Commission says it is true because judges will look at someone's criminal record in the sentencing phase and it turns out the average black criminal has a longer record than the average white criminal. I'm sorry! So Eric Holder acknowledged that blatant instances of racism where the people are looking at it, like what's going on in Minneapolis (I'm not saying it was race-based), it is an issue that will be analyzed, scrutinized, injected, respected, inspected, and it won't go away. I remember some years ago some executives from Texaco were on the phone making disparaging comments about black employees. Lost their jobs. Papa John, the CEO, on a phone call makes a disparaging comment about jobs. Loses his position at CEO.

This is what I'm talking about: *zero tolerance* for racism against blacks. Doesn't say it doesn't occur, I'm just saying there's zero tolerance for it. And that is one of the reasons, I'm sure, why (as I've said many times) studies have shown officers are more hesitant, more reluctant to use deadly force on blacks than on whites, even on Hispanics. Very likely because they know a ton of grief is going to come down on them. The civil rights establishment. The local media.

* * *

"Live, Liberty & Levin—Sunday, June 14"
Segment One: Carol Swain, PhD.
Segment Two: Wilfred Reilly, PhD
https://video.foxnews.com/v/6164156362001#sp=show-clips

Mark Levin: Hello, America. I'm Mark Levin. This is "Life, Liberty, & Levin." We have two wonderful guests tonight on part 2 of "The Insurrection."

Our first guest is **Dr. Carol Swain**. Dr. Swain taught at Princeton, where she was tenured. She was a tenured Professor of Political Science at Vanderbilt University. She has a B.A. in criminal justice, a Master's in law, a PhD in political science. Really a remarkable woman. She's also part of the "1776 Unite" group that Bob Woodson has organized and a "Black Voices for Trump" Advisory Board Member. Dr. Swain, you're very, very impressive. For people who don't know, give us a little bit of your background.

Carol Swain: Well, I'm an American success story. I was one of twelve children, born and raised in the rural South. We all dropped out of school after completing the 8th grade. I married at 16, had my first child by 17. By the time I was 20 I had three small children and I divorced, went to a community college, got the first of five degrees, and I eventually became a successful university professor. I earned early tenure at Princeton, I've won national prizes, I've been cited by the US Supreme Court, and I love America. And if you haven't noticed, I'm black.

Levin: Well, let me ask you some questions. That is fantastic background. You see what's been going on in this country. You've seen and heard the voices on television—athletes, the commissioner of the NFL, Hollywood the Democrat Party on the left, Pelosi and Schumer. First of all, I want to ask you a question. Is this a civil rights movement like the civil rights movement in the 1960s that had a specific agenda on equality, voting, access, due process? This seems to be an "insurrection" that has as its purpose the overthrow of the original principles of our government and the overthrow of our economic system. What do you make of this?

Swain: I think it's part of the cultural Marxist agenda against America and many of the protestors that are out there, the corporations that are giving them money, I believe they are sincere, that they believe that racism in America is intolerable, that it hasn't improved, and that black people live their lives in fear. That's a false narrative and I think it's being pushed this year because of the election. I think that with the situation with George Floyd, the media, the Democratic party, Antifa, Black Lives Matter, all of those groups have an incentive to push and fan the flames so that they ended up with what we're experiencing right now. We are all suffering as a result of their political agenda and it will not bring us closer together.

Levin: I look at this and I have to say, look at these cities, they've been run by one party with an iron fist—mayors, city councils, aldermen—for half a century at least. They're in charge of the money that comes into these communities; they're in charge of how that money is distributed. That money is laundered through their bureaucracy, their nepotism, their political cronyism. They take no responsibility for what's going on in their cities, they blame other people, they blame systemic racism, same with these one-party, Democrat states. It is amazing to me because you have Nancy Pelosi this week who announced that she wants to remove 11 Confederate statues from Congress, and I'm thinking to myself, "hasn't she been there like 30 years? She's been speaker twice, she was there when the Democrats controlled the House and the Senate and Barack Obama was President, and only now she decides to remove 11 Confederate statues?" Number one, what is that going to do for the people in these communities. And number two, isn't it, for her, purely opportunism, because she had every opportunity to remove those statues years ago!

Swain: Well, first of all I'd like for your viewers to know that I was a Democrat up until 2009, but I always had a different perspective on the world. I've always been able to see through things and so I've known that the Democratic Party was problematic, but I was there and I've been able to watch them over the years and it's clear that this whole movement to tear down monuments, I believe, it's really focused on dividing the country even further. The monuments are part of our history. If we can't learn from our history, then we are in trouble. The whole situation is problematic because there are monuments on all sides. If we start tearing down monuments, what happens tomorrow when the leaders that we are celebrating today fall out of favor? It's the wrong path for our nation. It's divisive. It's opportunism. It's just meant to stir up people that are traditionalists, people that love America, people that are patriotic.

Levin: I think you're right. The notion that we're going to tear down all the Confederate statues because Americans worship these generals? Americans don't worship these generals. Americans mostly despise these generals or don't even know who the hell they are. The fact is, you're going to start pulling down monuments and what comes next? Any of the founding fathers who owned slaves and released slaves? What comes next? Books? We already know movies are being affected, television shows are being pulled off the air. We have this kind of bizarre purging going on, this "cleaning up" of history going on, when in fact we actually need to know all these things as we go into the future. Correct?

Swain: It's far more important to look at the full picture and not to judge the past by the politically correct lens that we use today.

Levin: If we judge by the past by the "political lens of today," as you point out, then we have a lot of questions to ask of the left and the Democrat Party, don't we?

Swain: Yes.

Levin: For instance, how could you have voted for Robert Byrd to be the leader of the Democrat Party Majority Leader in the United States Senate when he had been an avowed Klansman several decades before? That seems to me to be a rational question. Or how could Joe Biden have worked with out-of-the-closet segregationists in the 1970s on various crime bills and other issues? If we're supposed to "cleanse" history, are we supposed to ignore that, too? Or may I say this to you, professor—the Democrat Party. If we're going to purge society of all vestiges of slavery, segregation, and racism, was it not the Democrat Party that led the Confederacy? Was it not the Democrat Party that gave birth to the Klan? Was it not the Democrat Party that throughout 100 years after the Civil War led the effort on segregation, and Jim Crow, and poll taxes, and literacy tests? I understand the Democrat Party has done some good things, but if it's in our DNA and we're systemically racist and we're pulling down statues, and we're burning books, and we're removing TV shows, why don't we purge the Democrat Party? Just a rhetorical point to you, professor.

Swain: I agree and the Democratic Party also fought against the anti-lynching laws. If you look at the 1960s, the Democrats have been manipulating black people for a long time. They pay off the leaders, they pick certain ones in the black community, they pay them off and nothing good trickles down to the masses. The Democratic Party has stood in the way of black progress. It's standing in the way right now. They are using black people to advance a radical agenda that I believe will be destructive to our nation, that it will hurt all of us, and it's hurting us right now.

Levin: When I look at relatively recent history, I look at 1964. The 1964 Civil Rights Act, which was actually quite crucial. You look at the breakdown of the votes; in the House of Representatives, 61% of Democrats voted for it, 80% of Republicans voted for it. In the Senate, where it was filibustered, 69% of Democrats wound up voting for it, 82% of Republicans wound up voting for it. And yet, the Democrat Party today claims to take the lead in dealing with what it calls systemic racism in society, systemic racism in police departments, that the Democrat Party is going to advance equality. You have iconic figures in our culture, Oprah Winfrey, LeBron James, you have the commissioner of the

NFL pretty much, these are all Democrats, all promoting the Democrat Party. Let me circle back. Is this, in many respects, about power and not so much about civil rights?

Swain: It's absolutely about power. If you go to the Black Lives Matter website and you click on the link to donate, that money is going to Democratic candidates. It's not really advancing the cause of black people. I think that it's very problematic that the true statement, the "slogan" that "black lives matter," in the same way that all lives matter, that because the slogan "black lives matter" is a true statement people that want to help blacks and they're white, they're being shamed, they feel like the only way they can show support is to donate money or to acquiesce or get down on their knees before black people, and that's not going to benefit the black community. And this whole move to defund the police, I mean that's ludicrous. That will hurt black people more than any other group.

Levin: I want to ask you a few questions. You're very experienced. You've lived what people like to theorize about. You are an example of the American dream. When you hear of people who haven't lived the way you've lived (the media, sportscasters, Hollywood), talk about a "systemically racist society," and so many of them do not live in these communities, maybe some of them came from these communities, they're in the top 1/100th percent of the income group, they live in these communities that are "systemically racist," they send their kids to schools in these communities that are "systemically racist." I'm talking about media figures, sports figures, and so forth. Does that kind of hypocrisy, again somebody like you who's come out of an inner city environment, does that kind of hypocrisy frustrate you?

Swain: Well the poverty that I came out of was a rural poverty, and what I've noticed as a student and as a faculty member is that a lot of people who have been privileged, whether they are white, black, or Hispanic, the more privileged they are, the more they feel the need to use that language of "systemic racism." I think it's being driven by guilt. They feel guilty because they're prospering. It's very problematic to me that the people who make public policy, a lot of them have degrees from Ivy League schools, they went to Ivy League schools and prep schools. They know nothing about the topics they're trying to solve. Maybe they did research, they were in a PhD program, they wrote a dissertation—that's the extent of their knowledge, and they're going to make public policy for the nation? That's one of the reasons that we're in trouble. All these theories that they put forth, the critical theory, critical feminist theory, critical race theory, all that's rooted in Marxism and it's a false narrative and it has no relationship to truth or reality.

Levin: This word Marxism. The fact is, the Antifa group, which has done enormous damage to inner city communities over the last couple of weeks, is a Marxist-anarchist group. The media pretends they don't exist and they downplay their role. Black Lives Matter, in many respects, is a very violent group and an anti-Semitic group, at least during its founding, part of the BDS Movement, that's never mentioned. People say they "stand with black lives matter, we stand for the lives of black people and all people. We do not necessarily (the world, certainly not I) stand with this organization, per se." Let me ask you this. School choice—education is key. Education has been key throughout your life. You're a highly-educated woman. In our inner cities, the kind of choice that we have in the rest of the country does not exist. We have an all-powerful labor union, a public-sector union, the National Education Association and the American Federation of Teachers. They are tied at the hip with the Democrat Party. They are huge donors to the Democrat Party, nearly exclusively, and they work hand-in-glove. Barack Obama was one of the most outspoken and ardent opponents of school choice which would allow the parents of poor kids, especially in violent communities or in communities with bad schools, and minority communities, to choose working, quality schools to send their kids rather than the school down the street if it's not functioning properly, and allows competition. Parochial schools, private schools, charter schools, and so forth, where the money follows the student, just as the money follows the student in college. This, to me, is a huge human and civil rights issue: basic education. The NBA is not behind it, the NFL is not behind it, the media are not behind it, Hollywood is not behind it, the Democratic Party (like so often in our history) is obstructing it. They talk about the cops' unions, what about the teachers' unions? What's your take on this?

Swain: I can tell you that in the public opinion polls that I've seen, majorities of blacks and Hispanics want school choice. It makes a lot of sense because the people who are making decisions for them have school choice, and in those inner-city schools they're crime-ridden, students are afraid to go to school because the "restorative justice" (rooted in Marxism) is making it almost impossible for students to be suspended. In cities that are run by Democrats, they have not gotten rid of that, so the lawlessness within the school house, the teachers are afraid to teach, there's chaos in schools, just like it is on the streets. I think that that's very problematic and we need school choice. We need to hold public schools accountable. Whenever Democratic cities want money they say "we have to pay our teachers more." They use the teachers, but a lot of that money never reaches the teachers. It's used for administrative purposes.

Levin: The problem is that these Democrat cities have an economic model that is more socialist than capitalist.

Swain: Yes.

Levin: They have an all-powerful bureaucracy and they do not allow these kids to get a proper education. Money keeps flowing through the same roots and it does not work. You want true revolutionary change? True, drastic reform? You've gotta remove these barriers—one-party rule of the public-sector teachers' union, and bring in more liberty, and opportunity, and prosperity, and education, and choice in the communities that haven't seen it in half a century.

[17:33—begin second segment]

Levin: Our next guest is **Dr. Wilfred Reilly,** Associate Professor of Political Science at Kentucky State University.

He has written a fantastic book, among others, *Hate Crime Hoax: How the Left is Selling a Fake Race War.*

Professor, you see what's going on around you. You've been commenting on this, so let me get right to the crux of the matter. Is there systemic racism in our police departments, in the aggregate?

Reilly: **Systemic, or institutional, racism, as an academic, is a phrase that I'm always a bit skeptical of.**

Now, obviously, if we want to be honest about some of this country's history there have been systems, like criminal sentencing in the past, where until quite recently you did see discrimination at kind of a broad, group-targeted level. But, very often, this phrase simply means there's a difference in performance between two groups and we're going to attribute that to racism.

To give an example of why that doesn't necessarily work, my favorite sports league, the NBA, the National Basketball Association, is more than 65% black, and I don't think any serious person would believe that that's because white jocks just don't get a fair shake in American society. The reason is that there's what you might call a "cultural variable." African Americans play basketball more, and so, on average at the median, with all respect to the hick from French Lick, we're better at it.

White guys play more baseball, and if you look at the MLB that's a very heavily Caucasian and Hispanic athletic league.

And you see this same thing in more, what you might call, "serious" situations all the time.

In 1995, the government economist June O'Neill, who's a liberal as I understand, and the conservative researcher Dinesh D'Souza, who was fairly nonpartisan at the time, looked at the troubling fact that there was a gap between black and white incomes.

As I recall, African Americans made 82% of what Caucasians did, and this is universally attributed to racism. What they found out was that this wasn't the case.

There were a lot of "variables" involved. African Americans are a younger population. The most common age for a black man is 27. The most common age for a white man is 58. Obviously, people earn more later in life, when they've had the chance to move into executive roles.

African Americans are much more concentrated in the South, where wages are lower for everyone. If you adjust for these factors and just a couple of others—years, not even quality of, schooling, just staying in school; average performance on a test like the SAT—the gap almost disappears. And you could argue that some of these things are tied into class and whatnot, but *there's no pattern of businesses paying an absolutely equally qualified black guy and an absolutely equally qualified white guy different amounts, in general.* And I think that's a valuable lesson to take into life.

So when people argue that policing is institutionally racist, almost always what they're saying is "African Americans are arrested at a higher rate," or "they encounter police at a higher rate" than the black percentage of society would predict. But the obvious intermediary variable there is crime rate.

If you look at the Bureau of Justice Statistics Crime

Report, the African American crime rate for violent crimes where you encounter the police (whites dominate corporate crime) is 2.4 times the white rate. So you would expect there, unfortunately, to be more encounters between African Americans and the police. And when you look at this narrative about black people and the police being in constant conflict, not only does a lot of the structural element (perhaps not all of it, but a great deal) disappear if you adjust for crime rate, the figures themselves that are often used strike me as very, very inflated in terms of police violence.

There's a comprehensive database of police shootings and police killings maintained by the *Washington Post* newspaper. There's another one that's kept online and live at www.killedbypolice.net.

If you look at that Washington Post database, for some reason the figures change from time-to-time, but until about a week ago, the total number of unarmed men killed by the police in the USA last year was listed as "9 unarmed black individuals." Total number of unarmed people (all races, all sexes) was only 56. There were only about 1,000 people of all backgrounds, armed or unarmed, killed by police, and only 229 of those were African American. There's a lot of serious research that looks at types of violence other than just shooting, other than killing. Roland Fryer, who is one of the youngest individuals ever to gain tenure at Harvard and, like me, is a Black man, took a comprehensive look at every type of police violence in a paper that came out in, I believe, 2016. What he found is that there were some small gaps. African Americans were a bit more likely to be, for example, cursed at or lightly shoved, but when it came to extreme beatings or police shootings, *African Americans in the case of shootings were actually 24.2% less likely, with everything along with race adjusted for, to be shot by the police than whites were.*

I don't think you see a pattern of broad, systemic targeting specifically of black people by the police today.

That doesn't mean there aren't individual bad officers. I mean, we saw the horrifying George Floyd tape. My position is that charges should be filed against every one of the officers involved in that situation. But even there, the facts are a little more complex than they originally seemed to be. Two of the officers on the scene, as it turns out, Thao and Kueng, if I have the names correct, are not Caucasian men. Floyd and Chauvin, the officer primarily responsible for the killing, had previously known one another from a nightlife security position. So I don't think we can just look at interactions between African Americans and the police and say "what we have here is an epidemic pattern of racism," a "genocide" as attorney Crump calls it in a recent book.

Levin: **Based on what you're saying, based on what I'm reading, based on the data, police brutality is not the issue that is holding back people in these various communities. What is holding people back in these various communities**, and why is the narrative so heavily focused on police brutality?

Reilly: I think that's a very complex question that gets into a lot of things, like the *difference between historic and contemporary racism.*

First of all, one thing that I would say is that minority communities in the United States have gone through and come through a lot. The black community endured slavery, at least in the South, endured Jim Crow, the crack era, so on down the line.

You've actually seen some pretty steady closures of gaps between blacks and whites. Just looking at this casually, doing some research the other day, the black SAT rose from about [950] in the 1970s, to [1800] looking at College Board data just a couple years ago.

So the black story in America is not a story of tragedy, but when you look at things that are genuinely problematic at a large scale in the black community I think you're seeing legacies of racism in the past, i.e. you're looking at poverty, housing, and so on down the line. Even more to the point, to me, you're seeing *legacies of the 1960s welfare state.*

***Fatherlessness*, for example, is the biggest problem (and I'm not just going to blame black people here) among all groups in America.** The great economist Walter Williams in 1938 looked at rates of fatherlessness for whites under it was under 4%, African Americans it was 11%. Those same rates today are 35.4% for Caucasians, including Caucasian Hispanics, over 50%, close to 60% for Hispanics and Native Americans, 72% for African Americans.

Obviously, the presence of a father, in particularly a young man's life, is one of the biggest predictors of how well he's going to do in high school, how likely he is to go to jail, things like suicide (that's especially true for lower-income Caucasian Americans).

So I think there are very broad systemic problems in the black community, and in many American communities. I would say ***the collapse of the family and the things associated with it, small business community, social sector, probably the biggest one.***

Why is it so tempting to blame the ghosts of the past? I think to some extent two reasons—one, we're used to doing so, and two, there is broad social approval.

I mean, celebrities, LeBron James for example, know that if they say "the cause of every one of these problems, such as fatherlessness or the test score gap, is racism," one, that is going to be greeted very approvingly by a network of natural allies throughout society. And two, it will sound like something that most people are used to hearing.

Dealing with the *actual* problems in African American communities and other communities is going to take more than the eradication of racism. **It's worth noting, by the way, that the large majority of people, black and white, are not racist in any conventional sense today.**

We might look at the results on some complex implicit bias test and be a little worried for a bit, but if you look at actual classic questions, "would you vote for a qualified member of the other race for president if they ran with your political party?" the figure for that is 92% for African Americans and for Whites.

Well over 80% of people support interracial dating, marriage, love.

I don't think we can simply look at an issue like violence in cities and say "well that's due to racism."

One thing you have to point out is that many of the large cities in the United States of America have been controlled by Democratic, often minority-led, administrations for

decades. If there is a problem with the schools in Detroit, or with gun violence in the street you can't necessarily just say "well that's the white man."

Levin: Also on this interracial issue you raised, I looked at statistics based on a few years back, and it indicated that for new marriages, 17%, or almost 1 in 5, are interracial. How can you have systemic racism when 1 out of every 5 marriages are interracial, and the other statistics you provide?

I could be wrong about this; I'd like to pose this to you: **this notion of "systemic racism" means everything and it means nothing. It's an excuse for not addressing the real issues that exist in these communities and other communities.** You pointed out one-party rule, one-party ideology, one-party economic ideology. There's a battle, there are obstacles to things like **school choice** that exist in other parts of the country but not in the inner cities. The Democrat Party that opposes, as I said earlier in the show, Barack Obama led the battle against school choice in many of these communities.

If you're going to have truly revolutionary and radical ideas, you have to have competition politically, competition in schools, and opportunities.

I'd like your thought on that. How about a liberty agenda, a competition agenda? That's sort of the environment that exists in so many other parts of the country? It's never discussed. You never hear it discussed by people in the media or in Hollywood, but **what about a real liberty agenda?**

Reilly: If you're talking about for the black community, I would fully support that. One thing I will say, I think that more competition, economically and politically, would be an excellent thing for the African American community. **I would love to see the black community split our vote 50/50, because that would imply that both political parties might actually do something for us.**

What you see right now is that every four years, with a few exceptions, there are a few long-term serving black civil servants, but Democratic politicians come out of the woodwork during election years and make impassioned, "racialized" pitches to the black community. This is true no matter who the GOP candidate is.

When Mitt Romney, who's a centrist, anti-racist Republican, was running for president, Joe Biden's comment about him was "he's gonna put ya'll back in chains." Mitt Romney's not going to bring back slavery, no matter which political party you vote for; he's just not going to.

This year the exact same line has been used against Donald Trump. In political science, by the way, we find that the mainstream media covered most GOP presidents as they have covered Donald Trump. The reaction to him isn't a result of his sometimes over-the-top offensive language, it's also how Romney was discussed, it's how Bush was discussed. But in regard to Trump, Joe Biden said "you ain't black" to the radio host Charlamagne Tha God, if you would consider voting for the Donald while he's in office. So I don't think that kind of virtue signaling every four years does much.

Again, obviously racism is real, but today, simply put, the biggest predictor of how successful you become in this country is not residual prejudice around you. *The most successful groups in the country have a competition or a liberty agenda.* Nigerian Americans, Japanese Americans, East Indian Americans,

Lebanese Americans, these, not white or black Americans, are the wealthiest Americans. I would definitely like to see a lot of that agenda transitioned into deprived communities, whether poor white communities or African American communities, although clearly my focus is on African American communities. I'd like to see **enterprise zones**. I'd like to see **charter schools** as an option for striving minority parents. HBCUs, a good number of which are private, I'm proud to teach at one, do a fair amount of good in this arena, but that's after high school. I would like to see as many chances to competitively perform in the 'hood as there are to competitively perform in the suburbs, and **that would do much, much more good for poor people, especially poor black people, than the endless ghost hunting for new, more-refined forms of racism.**

Levin: But you do understand what that means? That means trying to figure out *how to defeat one-party rule in these cities*, and *how to defeat the teachers' unions in these cities,* and a whole bureaucracy and a whole group of surrogates that have grown out of these one-party rule cities. That's going to be very difficult.

Reilly: *Public-sector unions*, in general, to me, as a businessman and an academic, are problematic. *Private-sector unions* are great. With public-sector unions you don't really have two parties coming to the table. You have the union, made up of middle-class employees and executives in the public sector, and then you have the politicians, to whom these unions give sizable donations. Those are the two parties negotiating and those are not two distinct parties. By the way, in terms of real solutions to the issues of policing, a lot of those would involve tackling police unions. You sometimes hear that "Camden got rid of the police." No they didn't. What they did was break the police union, give the cops some basic training in related fields, like social work, and changed their name, but they have 70% boots on the ground in terms of what are in practice police officers than they used to.

So the more you can take on these public-sector unions to achieve things like actual body cameras in every department, a discussion about no-knock raids, removing these various complex forms of immunity from contracts perhaps (I'd like to hear officer's reactions to that), the better a relationship between police and the communities you will see.

[Stop at 33:48]

* * *

Back to the Author's voice henceforth...

So what are these (mostly) "Black" Conservative voices saying as a whole?

Wait for it...

The same thing "White" Conservative voices are saying!

Yes, that's right.

So the point – and it's a big one – is that *your skin color is not the determining factor in how you think.* It is your **upbringing**, your **friends** and which Politics you identify with – and ascribe to!

Hitting it home: RACE has nothing to do with it! Doesn't control your mind! Most especially in that there is NO RACE!

***Polarization and division in America* primarily stems from a misguided allegiance to "race" as reality, Mainstream Media, Politics and ill-defined Subversive Groups who seek to undermine all of us at every turn...to seed doubt, fear, division and chaos. Not *maybe* that's where it comes from...I submit to you that here we have found our *source* of collective pain and guilt!**

Being somewhat facetious now, but skin pigment can not actually determine how a person thinks, speaks or *votes* or anything else exactly the same as with ALL other people who share one's unique shade of pigment.

We must discard the sometimes pejorative terms "black" and "white" because that all too convenient shorthand has caused us to speak in "us versus them" terms for far too long!

In the words of Morgan Freeman: "I am going to stop calling you white, and I ask you to stop calling me black." Our specific shade of skin does not define us.

I already dispelled in the last chapter the notion that many people are either one – but instead **we are an INFINITE range of hues, colors and shades.** It's harder to demonize 7.5 billion skin shades, each could be absurdly classified as a unique group, to ostracize, demean, demonize and disenfranchise!

After reading Chapter 4, **The Myth of Race**, one should never again plead ignorance of this unassailable fact:

THERE IS NO SUCH THING AS RACE – is not now, never was and never will be!

CHAPTER 6

MODERN MYTHOLOGY: DEMIGODS (AND DEMONS)

"One cannot consent to crawl, when one's impulse is to soar!"

~ Helen Adams Keller

Wow! Can I begin here, with her?!

Helen Keller is a Demi-Goddess! [If you are uncomfortable with the term demigod, substitute it with the word "luminary" or "visionary" in your mind].

My personal admiration for her – and her gifted teacher **Anne Sullivan** – knows no bounds! Born deaf and blind, Helen was in a dark abyss until she was helped toward the light by her loving teacher, and went on to become an inspiration to the entire world!

If you don't know her or know her well, you owe it to yourself to watch the movie of her life: "*The Miracle Worker!*" starring the inimitable actresses Patty Duke and Anne Bancroft.

See...from the title alone, you can see that those who created the movie recognized her inherent demi-goddess nature! Obviously without a gifted, patient, persistent, loving teacher, she would never have realized her potential. Thank God she did!

Why do I use the term **demigod**?

Because the men and women of whom I will speak are obviously not gods, but neither *merely* flesh and blood mortal either.

They are indeed *larger than life*, living or not, legends in their own time, or in time – and as such, *they fill the vast space between man and God.* Therefore, they are best thought of as demigods...or luminaries, visionaries, leaders and role models.

I'll offer the names of several men and women deserving of this title – or alternatively, any title you feel comfortable with – **for those rare individuals who are fully actualized human beings, so much so that they inspire us to be better people ourselves, so much so that the force of their personality, their very presence, their life force, if you will, inspires. The term for that is *charisma*.**

Well, these people had charisma as a starting point.

It was what they did (or *you* will yet do) with their God-given Gift (or if truly blessed, gifts) and charisma that

***really* counts.**

And these names did that so well that if you have spent any portion of your life being educated, self-educated or a keen observer, or just sat and watched educational television programming – Discovery, History and National Geographic just for starters, then you will recognize most or all of them.

Despite them being rare, they are also plentiful because so many billions have lived and died over the millennia that there are far too many for me to mention:

I could write in a stream of consciousness state without pausing for hours if my forearms wouldn't saturate with lactic acid, the burn set in, and my stiff fingers would not wear down to the nubs…and still never list them all.

Most are **famous**. Some are **good** or even **Holy.**

Men and women like George Washington, Thomas Jefferson, Benjamin Franklin or any of our forefathers (and please, no politically correct, historical revisionism lest you be judged in kind), Abraham Lincoln. Men like Jesus Christ.

Whoa, stop right there! you think. Jesus is not merely a man; he's God or God's Son or the Holy Spirit or all three… the Holy Trinity. He's our Lord and my personal Savior!

Oh, I couldn't possibly agree more, but not to everyone, and so he would have to be at or near the top of *anyone's* list nonetheless – Christian or not. Even the historical Jesus easily makes it to the top of this list!

If you had the opposite reaction and don't think so,

please share with us any human being, alive or dead, that has influenced more human beings. Jesus is not just prophesied in the Bible's Old Testament and the central figure of the New Testament; he is even described in the Koran as a revered Prophet, so that adds billions more souls instantly. His reach is incalculable!

For Jews, we must include the Prophets of the Old Testament in this group. We are all descendants of Abraham (also lauded in the Koran). For Catholics, we need to include all the Saints.

I'll list just one contemporary Saint: Mother Teresa.

Choosing to spend her live caring for the bodies and spirits of the least fortunate, the poverty stricken and ill, she is the very definition of a compassionate human being. Yet even Mother Teresa had her detractors, but she remains my own definition of a saint, demi-goddess, luminary, visionary and leader / role model.

For our purposes here, we are speaking of flesh and blood human beings, warts and all. Being flawed in some way or guilty of having sinned is not a dis-qualifier, or we have no list at all! That's not very hopeful for humanity.

Let's repeat the definition: **Rare individuals who are fully actualized human beings, so much so that they inspire us to be better people ourselves, so much so that the force of their personality, their very presence, their life force, if you will, inspires us for good or moral action.**

Per our definition, we can eliminate larger-than-life individuals like Alexander the Great, Kublai and Ghenghis Khan, Napoleon, Stalin, Pol Pot, Hitler…because as either conquerors, dictators or megalomaniacal despots, they fail the criterion "to inspire us toward good or moral action," but any name is up for debate as a candidate in an academic setting.

Modern-day visionaries like Elon Musk would make my own list, especially after his Space X venture placed two American astronauts into low-earth orbit on the International Space Station, but possibly he would not make your list. Certainly it took more than Elon's leadership to ensure that success, but he was nonetheless the visionary behind the historical achievement.

Please feel free to make your own list.

As far as "**demons**" go, I made that clear at the end of the last chapter, and will merely repeat it here:

Polarization and division in America **primarily stems from Mainstream Media, Politics and ill-defined Subversive Groups (e.g. AntiFa, BLM, anarchists, arsonists and professional, financially-supported instigators) who seek to undermine all of us at every turn...to seed doubt, fear, division and chaos. Again, not maybe that's where it comes from...I submit to you that here we have found our source of collective pain and guilt!**

I do not wish to go into the leftist / socialist and Marxist foundation and leanings of these groups lest I stray too far from the central focus on transcending race altogether, but I assure you there is a factual basis for this assertion.

As I stated above...

A demigod / luminary / visionary / leader / role model

that I could actually admire and wish to emulate, by *my own definition*, would be:

A charismatic person with *any* form of God-given gift that actually uses it in the world in the service of others.

Remember this definition well, my dear Reader, because I will provide you an opportunity to become a demigod / luminary / visionary yourself – and no, that's not writing or poetic license, or even hyperbole.

Before the last word is typed in this book, I pledge to offer you a chance to embark on your own personal journey toward being one of those rare souls we all admire…at a minimum, to become a minor luminary who develops a small following of family, friends and others, perhaps hundreds, thousands or who knows, even millions of others.

In fact, minor demigods, luminaries or visionaries are what we all have the potential to be!

If you happen to be a Christian, you already accept that we are all made in God's image. I think that almost qualifies each of us for demigod status already, don't you?!

The only other parts of the definition you need to fulfill is to **discover the God-given gift you possess, to become charismatic, to influence others in a positive way or to then act in service to others…**

* * *

Martin Luther King, Jr.

An appropriate **"race-specific" luminary** who was certainly gifted, certainly charismatic [for goodness sake, his birthday is an American Holiday!] who was essentially good, although complex and flawed like all other mortals at the same time, and who used his charisma and gifts in the service of others!

One of MLK's gifts – as we all know – was his *oratorical skills.*

Who could forget his "**I have a Dream**" speech where **"Someday, all God's children will be judged by the content of their character...and not the color of their skin!"**

Other famous Quotations by Martin Luther King, Jr.—

From the sides of the hallowed granite stone monuments inside "MLK's Memorial" in Atlanta, Georgia—

In the summer of 1992, standing at the MLK Memorial, I read and was inspired by the "Wall of Quotes" spanning the years of Martin Luther King's long ministry and civil rights career...

My *favorite quote* is:

"We are determined (here in Montgomery) to work and fight until justice runs down like water. And righteousness like a mighty stream!"

In part it was my favorite quotation because of the vivid imagery it evokes, but also because I lived at the time of The Civil Rights Movement in Montgomery, Alabama, was fully steeped in the culture of the Civil Rights Era.

My mother and I together, but apart from the rest of our family, were just getting ready to move to Colorado Springs, Colorado. My sister had married an Air Force Academy graduate, left to make her new life in Northern California temporarily while her husband attended graduate school at Stanford. [Ironically, the Hoover Institute is there and some of the luminaries mentioned in this chapter are Fellows].

My brother was at The Citadel, The Military College of South Carolina, and my father was departing for South Vietnam as the Commander of a Squadron performing dangerous missions over the Ho Chi Minh Trail, the De-Militarized Zone between war-torn adversaries at the height of the War / Conflict (as it is alternatively known).

So it would be just the two of us, and in a way it was a departure from the "hot" South – both literally and figuratively due to racial tensions – to the cooler clime, both literally and figuratively, as Colorado Springs at that time, and even to this day is a military haven – and you will recall from Chapter One how I described that Montgomery, Alabama in 1964-1967 was like living inside the movie "The Help," yet the military bases were "race-free havens" for many people.

So returning to MLK quotations—
On his imposing sculptural statue:
"Out of the mountain of despair, a stone of hope."
"I was a drum major for justice, peace and righteousness."

The South Wall

"I oppose the War in Vietnam..."

Tying this into our topic, depending on your own politics, you might substitute Coronavirus, corrupt politicians, the politics and philosophy behind racial division, ANTIFA, BLM, race riots or any other pernicious or evil, malevolent force like "***The Beast***" in the movie "Poltergeist" (Chapter Two), or in our context within this book: the *false narratives of "systemic racism" and the social construct of "race."*

Despite the false social construct of "**race**" and therefore the associated false notion of "**systemic racism**," exacerbated by a plethora of corrupt politicians, don't lose sight of our progress. Don't lose sight of the role you can yet play in bringing about this "love gift" to the world: to not only **reject "race" as a concept and a construct**, but to:

TRANSCEND "RACE" ENTIRELY

To live on a higher plane of existence, and lead others to a more righteous path and day in which all God's children, being equal in the eyes of God, are equally so in the eyes of Mankind!

If we are to have PEACE on EARTH, good will toward men, our new found loyalties must...

TRANSCEND RACE...

CHAPTER 7

WHERE DO WE GO FROM HERE?

"The tribe often thinks the visionary has turned his back on them, when, in fact, the visionary has simply turned his face to the future!"

~ Ray Davis

"Our cause is noble; it is the cause of Mankind!"

~ George Washington

"The One Who Plants "Trees," knowing that he [or she] will never sit in their shade...has started to understand the meaning of life!"

~ Anonymous

Where do we go from here? Why, toward "*The Final Frontier*" obviously.

What is ***the final frontier***, you ask?

By now, you know deep down inside what it is, but perhaps can't yet articulate it.

My inspirational quotes at the top of each chapter were consciously intended to act as *metaphorical breadcrumbs* like Hansel and Gretel used to find their way out of the deep, dark, dangerous woods…

Recall this exceptionally prescient, prophetic one?

"There is a great new work before us, which is to replace with true knowledge the past ignorance that has destroyed human minds!

We will construct unity in a world which has been brutally torn apart by false divisions based on "race"…

We will *heal with unconditional love* those souls whose hearts have been disfigured by hatred and loneliness."

I now lift the veil fully from your eyes and reveal to you that:

> **Unity…achieved through unconditional love, healing our collective hearts, minds and souls is our metaphorical final destination on the journey of self-discovery, on the path to knowledge—our "final frontier!"**

But please, at least need some guidance, "real" steps or actions I should strive to take "on this path less traveled, less worn!"

Absolutely! You invested the time to read this book, and I will provide the promised payoff.

Within the context of this book – and as it pertains to so-called "race..."

Here are the steps / actions you can choose to take to move forward in the intended spirit:

1. First, you must *discover your own God-given gift* that every single one of us possesses as a child of God. [My own Gift (or Super Power, if you will), people tell me, is that I am a good teacher and communicator, especially via *the written word.*]
2. Become "charismatic" through self-examination, by listening to others, by exhibiting real compassion, by shining a "light" on others rather than acting vainglorious. Seek gradually to increase your influence – but acting strictly for the good of mankind!
3. Act within yourself initially, but then project externally into the world, using your own gift(s) to make the world a better place!
4. Become the demigod, luminary, visionary, leader or role model you have always wanted to be in your own life – and in the lives of others.
5. Reject entirely the false notion of "**race**." Promise yourself *to never use the divisive nature of "race" again as your personal reality. [*All who have come to the end of their journey of knowledge, as I very nearly have myself if only because of advancing age, who have now abandoned and tossed "ignorance" into the dustbin

of history where it belongs – *must* acknowledge the demise of this outdated, outmoded, divisive and evil notion to be able to take the final steps together with increasing confidence.]

6. Make the world a better place for yourself, your loved ones, your local community, your country and for the world!
7. Leave a *legacy of love* behind when you depart this world for the next plane of existence.
8. *Death of the corruptible mortal body (corpus) is real.* Death is a rapid transition from this mortal coil to a heavenly plane where your Celestial Body is no longer corruptible, no longer failing you, no longer breaking down, getting sick, wasting away, rotting in the end.
9. *Death of one's existence, one's soul, is an illusion* – and a thin one at that, just like the "veil" we have been referencing throughout this book that obscures your vision – which should come as a great relief to you!
10. *This illusion that "death" is real rather than the false construct it actually is (just as with "race")* is the final veil you must lift from your eyes, and in so doing, to banish all fear, the fear you have felt almost every waking day of your life, and in so doing, actually *transcend...*

"Death and Beyond:
THE ULTIMATE FRONTIER!"

You guessed it, dear Reader! That's one of my *next* books!

Look for—

Other Books by **Christian Veritas**
(Released by Peregrine Publius Publishing):

“RACE”: The Final Frontier!

- Publication date set for July 2020

“LIFE EXTENSION to IMMORTALITY”: The Expanding Frontier!

- Christmas 2020

“DEATH?”: The Ultimate Frontier!

- Spring 2021

“FUSION: The Infinite Frontier!

- Christmas 2021

Made in the USA
Monee, IL
04 November 2026

67380053R00085